Prentice Hall

WORD PROBLEM WORKBOOK

for

DEVELOPMENTAL MATHEMATICS

KELLI JADE HAMMER

Broward Community College

Upper Saddle River, NJ 07458

Editor-in-Chief: Chris Hoag
Senior Acquisitions Editor: Paul Murphy
Supplement Editor: Elizabeth Covello
Assistant Managing Editor: John Matthews
Production Editor: Allyson Kloss
Supplement Cover Manager: Paul Gourhan
Supplement Cover Designer: Joanne Alexandris
Manufacturing Buyer: Ilene Kahn

© 2004 Pearson Education, Inc.
Pearson Prentice Hall
Pearson Education, Inc.
Upper Saddle River, NJ 07458

All rights reserved. No part of this book may be reproduced in any form or by any means, without permission in writing from the publisher.

Pearson Prentice Hall® is a trademark of Pearson Education, Inc.

The author and publisher of this book have used their best efforts in preparing this book. These efforts include the development, research, and testing of the theories and programs to determine their effectiveness. The author and publisher make no warranty of any kind, expressed or implied, with regard to these programs or the documentation contained in this book. The author and publisher shall not be liable in any event for incidental or consequential damages in connection with, or arising out of, the furnishing, performance, or use of these programs.

Printed in the United States of America

10 9 8 7 6 5 4

ISBN 0-13-144788-2

Pearson Education Ltd., *London*
Pearson Education Australia Pty. Ltd., *Sydney*
Pearson Education Singapore, Pte. Ltd.
Pearson Education North Asia Ltd., *Hong Kong*
Pearson Education Canada, Inc., *Toronto*
Pearson Educación de Mexico, S.A. de C.V.
Pearson Education—Japan, *Tokyo*
Pearson Education Malaysia, Pte. Ltd.

Table of Contents

Section 1	Direct Translation	Page 1
Section 2	Basic Percents	Page 8
Section 3	Proportions	Page 14
Section 4	Parts	Page 19
Section 5	Geometric Perimeter	Page 27
Section 6	Consecutive Integers	Page 35
Section 7	Motion – Linear Equations	Page 43
Section 8	Money	Page 51
Section 9	Simple Interest	Page 61
Section 10	Mixture	Page 70
Section 11	Geometric Area and the Pythagorean Theorem	Page 80
Section 12	Motion – Rational Equations	Page 87
Section 13	Work	Page 96

Section 1

Direct Translation

Mathematics has a language all its own. In order to be able to solve many types of word problems, we need to be able to translate the English Language into Math Language.

Direct Translation is the process of translating English words and phrases into numbers, mathematical symbols, expressions, and equations.

Translating words into math is the foundation to understanding and successfully solving word problems. In this section, you will learn your new Math Language. Learning this language will make solving word problems easier than you would have ever imagined.

- **Basic Operation Words** *Sum, Difference, Product, Quotient*

Basic Operation Words tell you what operation to perform. An operation refers to a mathematical calculation such as adding, multiplying, or using exponents.

Basic Operation Words are always accompanied by the word "and". The operation word tells you what operation to perform and which symbol to use; the word "and" tells you where to put the symbol.

Basic Operation Words	What They Mean	Symbol Used In Place Of "And"
The sum of	To add	+
The difference of/between	To subtract	−
The product of	To multiply	· or ( )
The quotient of	To divide	÷

For example, in a word problem using the phrase: **the *sum* of 7 *and* 4**, the word "sum" tells you that you are using addition. The word "and" tells you where to put the symbol of addition, the plus sign (+). In this case, the phrase "**the sum of 7 and 4**" in English would translate to "$7 + 4$" in Math Language.

    TRANSALATION EXAMPLES

        English: The difference between x and 5   **English:** The quotient of 16 and r
        Math: $x - 5$ **Math:** $16 \div r$

- **Reverse Operation Words** *More Than, Added To, Less Than, Subtracted From*

Reverse Operation Words also tell you what operation to perform and which symbol to use. But Reverse Operation Words *reverse* the order of the numbers, variables, or expressions.

Reverse Operation Words	What They Mean	Symbol Used With Reversed Order Of Numbers, Variables, Or Expressions
More than	To add	+
Added to	To add	+
Less than	To subtract	−
Subtracted from	To subtract	−

For example, in a word problem using the phrase: **6 *less than* 8**, the phrase "less than" tells you that you are using subtraction so you will use a minus sign. But also, because "less than" is a Reverse Word phrase, you need to change the order of the numbers. So in this case, the English phrase "**6 less than 8**" would translate to "**8 − 6**" in Math Language.

TRANSLATION EXAMPLES

English: y added to 10
Math: $10 + y$

English: 6 subtracted from x
Math: $x - 6$

- **Miscellaneous Operation Words** *Increased By, Decreased By, Twice, Squared*

These operation words are not accompanied by the word "and". The Miscellaneous Operation Words will tell you what operation to perform, which symbol to use, and how a specific English word translates into Math Language. When using Miscellaneous Operation Words, the order of the numbers, variables, or expressions remain the same.

Miscellaneous Operation Words	What They Mean	How To Translate -- □ Represents Number, Variable, Or Expression
Increased by	To add	+
Decreased by	To subtract	−
Twice	Two times the number	2 · □
Squared	To use an exponent of 2	□²
Cubed	To use an exponent of 3	□³

TRANSLATION EXAMPLES

English: 14 decreased by y
Math: $14 - y$

English: Twice m
Math: $2 \cdot m$

English: w squared
Math: w^2

- **Miscellaneous Translation Words** *Is, The Result Is, A Number*

These words are the missing links that will help you fully translate English sentences into Math Language. Translation Words do not give you a specific operation to perform, but they must be translated in order for you to change your word problem into a math equation you can solve.

Miscellaneous Translation Words	What They Mean	How To Translate
Is	Is equal to	=
The result is	Is equal to	=
A number	An unknown (variable)	x or n or any letter

TRANSLATION EXAMPLES

English: Four times five is twenty
Math: $4 \cdot 5 = 20$

English: Seven decreased by a number
Math: $7 - n$

- **Important Multiplication Translations Rules**

Sometimes in a word problem you will see a phrase that tells you to multiply an amount times a number, such as "**three times a number**". This means you would multiply three times a single item. In this case, it would be 3 times the variable x, or in Math, "**$3x$**".

However, when you see a phrase that tells you to multiply an amount times an *operation*, such as "**three times the sum of y and 5**", this means you would be multiplying three times the *result of the operation*.

In this case, you would first have to translate the operation words into Math Language, and then put that resulting expression in parentheses to be multiplied by 3, such as "**$3(y + 5)$**".

This pattern would hold true no matter what type of operation it would be. The examples below illustrate this important concept.

TRANSLATION EXAMPLES

English: Five times the difference of a number and three
Math: $5(x - 3)$

English: Six times a number
Math: $6n$

- **Putting It All Together**

With the information you now have, it is time to put everything together. By being able to translate English word sentences into Math Language, the word problems will become normal math equations.

HELPFUL HINTS

- Do not look at the problem as one long sentence. Look for keywords and break down the sentence into separate phrases.

- The first thing to look for is a word or phrase that indicates where the equal sign will be.

- Whatever words are in front of the phrase that means "is equal to" will translate into the math expression that goes before the equal sign. Whatever is after that phrase will translate into the expression that goes after the equal sign.

- Be careful to look out for Reverse Operation Words so you remember to reverse the order of the number and variables when you translate from English into Math.

- You may come across some Operation Words that are not in the charts. For example, "shorter than" would mean the same as "less than". "Higher than" may mean the same as "more than". Use common sense to help figure out the meaning of a phrase.

Word Problem Workbook – Direct Translation Table

Key Words	What They Mean	How To Translate
The sum of	To add	+
The difference of/between	To subtract	−
The product of	To multiply	• or ( )
The quotient of	To divide	÷
More than	To add	+
Added to	To add	+
Less than	To subtract	−
Subtracted from	To subtract	−
Increased by	To add	+
Decreased by	To subtract	−
Twice	Two times the number	$2 \cdot \square$
Squared	To use an exponent of 2	$\square^2$
Cubed	To use an exponent of 3	$\square^3$
Is	Is equal to	=
The result is	Is equal to	=
A number	An unknown (variable)	x or n or any letter

Direct Translation   Word Problem Workbook

EXAMPLES

EXAMPLE 1 Six subtracted from a number is -5.
SOLUTION

English – Bold Type	Math Translation
Six subtracted from a number **is** -5.	Six subtracted from a number $= -5$
Six subtracted from **a number** is -5.	Six subtracted from $x = -5$.
Six **subtracted from** a number is -5.	$x - 6 = -5$

EXAMPLE 2 If 7 is added to 3 times a number, the result is the difference of the number squared and 8.
SOLUTION

English – Bold Type	Math Translation
If 7 is added to 3 times a number, **the result is** the difference of the number squared and 8.	If 7 is added to 3 times a number, $=$ the difference of the number squared and 8.
If 7 is added to **3 times a number**, the result is the difference of **the number squared** and 8.	7 is added to $3x =$ the difference of x^2 and 8.
If 7 **is added to** 3 times a number, the result is **the difference of** the number squared **and** 8.	$3x + 7 = x^2 - 8$

EXAMPLE 3 Twice the sum of a number and 6 is equal to one more than -10 times the number.
SOLUTION

English – Bold Type	Math Translation
Twice the sum of a number and 6 **is equal to** one more than -10 times the number.	Twice the sum of a number and 6 $=$ one more than -10 times the number
Twice the sum of **a number** and 6 is equal to one more than -10 times **the number**.	Twice the sum of x and 6 $=$ one more than -10 times x
Twice the sum of a number and 6 is equal to one more than -10 **times** the number.	2(the sum of x and 6) $=$ one more than $-10x$
Twice **the sum of** a number **and** 6 is equal to one **more than** -10 times the number.	$2(x + 6) = -10x + 1$

Word Problem Workbook   *Direct Translation*

NOTE: *Translate only. Do not solve the equations.*

Section 1: Direct Translation Exercise Set

1. The sum of 12 and a number is 10.
2. The difference between a number and 6 is − 4.
3. Twice a number decreased by 2 is 11.
4. Three times a number increased by 7 is − 1.
5. Six less than a number cubed is 15.
6. A number squared added to 8 is − 5.
7. Five times the sum of 6 and a number is 24.
8. Twice the sum of a number and 10 is − 20.
9. If 3 is subtracted from 4 times a number, the result is the sum of 5 times the number and 10.
10. If a number is added to 6, the result is the difference between twice the number and 5.
11. Four times the sum of a number and 10 is − 92.
12. Six times the difference of 7 and a number is 12.
13. If a number is subtracted from 94, the result is 19 more than the product of 5 and the number.
14. If 3 is added to twice a number, the result is the difference of 4 times the number and 5.
15. Eight subtracted from 9 times a number is equal to the sum of 6 and the number squared.
16. The sum of 12 and a number is equal to the number cubed added to 5.
17. Eleven times the difference of a number and 42 is equal to the sum of the number cubed and 10.
18. The product of 6 and a number added to 3 is equal to the number squared increased by 8.
19. If 7 is subtracted from a number squared, the result is twice the sum of 5 times the number and 2.
20. Six times the difference of twice a number and one is equal to the number cubed less than 16.

Section 2

Basic Percents

Basic Percents are word problems where you are asked to find a number or a percent. These word problems are very basic when it comes to translating English Language into Math Language.

- **Direct Translation Words For Basic Percents** *Is, Of*

When translating from English to Math in Basic Percent word problems, the words "**is**" and "**of**" translate directly into math symbols. As mentioned in Section 1, "**is**" translates to "**equals**". The Direction Translation Word "**of**" translates to multiplication.

> (*Note*: It is recommended that you use the times sign (×). to indicate multiplication. In Basic Percent word problems you will almost always be working with decimals. You want to avoid confusing the decimal point with the multiplication dot.)

Direct Translation Words	What They Mean	Symbol Used
Is	Is equal to	=
Of	Multiply	×

TRANSLATION EXAMPLES

English: 14 is what percent of 80?
Math: 14 = what percent × 80?

English: What is 5% of 24?
Math: What = 5% × 24?

- **Direct Translation Word** *What*

The Direct Translation Word "what" in Basic Percents will either be accompanied by the word "**number**" or the word "**percent**". It doesn't matter with which word "**what**" is partnered. "**What**" always represents an unknown, and is therefore replaced with a variable.

> (*Note*: It is recommended that you use the letter "n" as the variable instead of "x". With Basic Percent word problems, you will be working with the times sign (×). You want to avoid confusing the variable "x" with the times sign..)

Word Problem Workbook   *Basic Percents*

- **Solving The Problem** *Set Up And Solve The Equation, Answer The Question Asked*

The following steps will help you solve Basic Percent word problems:

Step 1
Set Up The Equation Using Direct Translation

When setting up an equation for a Basic Percent Word Problem, translate the English sentence in order from left to right.

- When you see the word "**is**", replace it with an equal sign.
- When you see the word "**of**", replace it with a times sign.
- When you see the words "**what number/percent**", replace them with a variable.
- If you see a number, just re-write the number in your Math equation.

  (*Note*: You must change the percent to a decimal or fraction before writing it in your equation. In this chapter, all percents will be changed to decimals. For example, "**20%**" would be changed to "**0.20**" or "**0.2**".)

Step 2
Solve The Equation

Using the method taught by your instructor, solve the equation for the variable.

  (*Note*: If necessary, you may want to simplify the appearance of the equation to a more familiar format. For example, if your equation states "$n \times 15$", it would probably be more familiar and recognizable to you to change it to "$15n$".)

Step 3
Make Sure You Answer The Question Asked

Once you solve the equation, you will find the value for the variable "n", but that might not be the answer to the question. In Basic Percent word problems, the question will ask:

    "What?"
    "What number"
    "What percent?"

If the questions asks "What" or "What number," the value of n is your final answer and you are done with the problem.

But if the questions asks for "What Percent", then the value of n is NOT the correct answer. You must change the value of n into a percent in order to have the correct answer.

Basic Percents Word Problem Workbook

EXAMPLES

EXAMPLE 1 What is 18% of 84?

SOLUTION

Step 1 *Set Up The Equation*

- Translate in order from left to right.
- "What" becomes a variable. "Is" becomes an equal sign. "Of" becmes a times sign.
- Change 18% percent into a decimal.

$$n = 0.18 \times 84$$

Step 2 *Solve The Equation*

- In this example, the equation is already in a familiar form.
- Since the variable, n, is already by itself, there is no need to solve for n.
- To get the answer to your equation, multiply 0.18×84.

$$n = 15.12$$

Step 3 *Answer The Question Asked*

- The questions asks "What?", so the value of n IS your final answer and you are done.

Answer: 15.12

Word Problem Workbook   Basic Percents

EXAMPLE 2 51 is 5% of what number?

SOLUTION

Step 1 *Set Up The Equation*

- Translate in order from left to right.
- Change the Basic Percent Translations Words into their symbols.
- "Is" translates to "=". "Of" translates to the times sign. "What number" is the variable.
- Change 5% percent into a decimal.

$$51 = 0.05 \times n$$

Step 2 *Solve The Equation*

- You may want the equation in a more familiar form. Change "$0.05 \times n$" to "$0.05n$".
- If you prefer the variable to the left of the equal sign, it's okay to switch the expressions.
- Using the method taught by your instructor, solve the equation for n.

$$51 = 0.05n$$
$$0.05n = 51$$
$$n = 1020$$

Step 3 *Answer The Question Asked*

- The questions asks "What number?", so the value of n IS your final answer and you are done.

Answer: 1020

Basic Percents   Word Problem Workbook

EXAMPLE 3   25 is what percent of 200?

SOLUTION

Step 1 *Set Up The Equation*

- Translate in order from left to right.
- Change the Basic Percent Translations Words into their symbols.
- "Is" translates to "=". "What percent" is the variable. "Of" translates to the times sign.

$$25 = n \times 200$$

Step 2 *Solve The Equation*

- You may want the equation in a more familiar form. Change "$n \times 200$" to "$200n$".
- If you prefer the variable to the left of the equal sign, it's okay to switch the expressions.
- Using the method taught by your instructor, solve the equation for n.

$$25 = 200n$$
$$200n = 25$$
$$n = \tfrac{1}{8} \text{ or } 0.125$$

Step 3 *Answer The Question Asked*

- The questions asks "What percent", so the value of n as given is NOT your final answer.
- Change the value of n to a percent to get the correct answer.

Answer: 12.5%

Section 2: Basic Percents Exercise Set

1. What is 5% of 80?
2. What is 20% of 150?
3. 21 is 3% of what number?
4. 3 is 15% of what number?
5. 40 is what percent of 200?
6. 16 is what percent of 120?
7. What is 16% of 75?
8. What is 8% of 275?
9. 18 is 30% of what number?
10. 4 is 5% of what number?
11. 11 is what percent of 50?
12. 100 is what percent of 300?
13. What is 6.5% of 74?
14. What is 21.4% of 85?
15. 12 is 120% of what number?
16. 5 is 200% of what number?
17. 75 is what percent of 15?
18. 84 is what percent of 21?
19. What is 120% of 15?
20. What is 220% of 3?

Section 3

Proportions

Proportions Word Problems involve setting up an equation with two equivalent fractions. In this section you will learn how to identify the pieces of information given in these types of problems. This will make them easier to set up and solve.

Step 1
Set Up A Proportion And Determine The Amounts And Their Categories

Begin by setting up a blank proportion, which is one fraction bar set equal to another fraction bar. In a Proportions Word Problem, you will always be given three pieces of information. The information given will be three different amounts that fall into two categories.

In order to determine what category the amount falls into, look at the noun that follows the amount. For example, if the problem states "**3 dogs**", "**3**" is the ***amount*** and "**dogs**" is the ***category***. Similarly, if it says "**8 months**", "**8**" is the amount, and "**months**" is the category.

Step 2
Fill In The Blanks Of The First Fraction

As you read through the problem, you will come across your first piece of information. This will be an amount with its category. Write this information in the *numerator* of the first fraction (the fraction to the left of the equal sign).

As you continue to read, you will come across the second piece of information which will be another amount with its category. Write this second piece of information in the *denominator* of the same first fraction.

Step 3
Fill In The Blanks Of The Second Fraction

The placement of the third piece of information **is key to the successful set up of the proportion**. The third piece of information is the final amount and category given in the word problem. The category of this third amount will match one of the categories you have already written in the first fraction.

- If the category of this third amount matches the category in the **numerator** of the first fraction, write this amount with its category in the ***numerator*** of the second fraction.

- If the category matches the category in the **denominator** of the first fraction, write this amount with its category in the ***denominator*** of the second fraction.

- Put a **question mark** in the remaining blank of the second fraction to indicate the question you are being asked in the word problem.

Word Problem Workbook　　　　　　　　　　　　　　　　　　　　　　Proportions

Step 4
Solve the Equation

Rewrite the proportion as a mathematical equation. Use the amounts only (no categories) and replace the question mark with a variable. Then solve the equation using the method taught by your instructor.

EXAMPLES

EXAMPLE 1 Tara can type 4 pages of her term paper in 30 minutes. How long will it take her to type the term paper if it has 14 pages?

SOLUTION

Step 1 Set up a blank proportion. The amounts and categories are: 4 pages, 30 minutes, 14 pages.	____ = ____
Step 2 Fill in the blanks of the first fraction. "4 pages" belongs in the numerator and "30 minutes" belongs in the denominator.	$\dfrac{4 \text{ pages}}{30 \text{ minutes}} = $ ____
Step 3 Fill in the second fraction. "14 pages" is the same category as the numerator of the first fraction, so make sure to place it in the numerator of the second fraction. Put a question mark in the missing blank.	$\dfrac{4 \text{ pages}}{30 \text{ minutes}} = \dfrac{14 \text{ pages}}{?}$
Step 4 Rewrite the equation with amounts only and replace the question mark with a variable. Solve the equation.	$\dfrac{4}{30} = \dfrac{14}{x}$ $x = 105$

Answer: It will take Tara 105 minutes.

Proportions   Word Problem Workbook

HELPFUL HINT

- You can't always determine the category by the one word that follows the amount. For example, a problem may say "4 cups of water" and "6 cups of juice". Be careful not to just see the word "cups" and assume both amounts fall into the same category. They do not: One category is "cups of **water**", while the other category is "cups of **juice**".

- Use common sense. The word following the amount is not always the category. For instance, if the question says "$8.00", 8 is the amount and $ (dollars) is the category.

EXAMPLE 2 If 3 ounces of medicine must be mixed with 5 ounces of water, how many ounces of medicine must be mixed with 15 ounces of water?

SOLUTION

Step 1	Set up a blank proportion. The amounts and categories are: 3 ounces of medicine, 5 ounces of water, and 15 ounces of water. To determine the correct categories, see *Step 1 Helpful Hint*..	____ = ____
Step 2	Fill in the blanks of the first fraction. "3 ounces of medicine" belongs in the numerator and "5 ounces of water" belongs in the denominator.	$\dfrac{3 \text{ oz. medicine}}{5 \text{ oz. water}} = $ ____
Step 3	Fill in the second fraction. "15 ounces of water" is the same category as the denominator of the first fraction, so make sure to place it in the denominator of the second fraction. Put a question mark in the missing blank.	$\dfrac{3 \text{ oz. medicine}}{5 \text{ oz. water}} = \dfrac{?}{15 \text{ oz. water}}$
Step 4	Rewrite the equation with amounts only and replace the question mark with a variable. Solve the equation	$\dfrac{3}{5} = \dfrac{x}{15}$ $x = 9$

Answer: 9 ounces of medicine.

Word Problem Workbook   Proportions

Section 3: Proportions Exercise Set

1. A concrete cement mixer uses 4 tanks of water to mix 15 bags of cement. How many tanks of water are needed to mix 30 bags of cement?

2. Katelyn drives her car 235 miles in 5 hours. At this rate how far will she travel in 7 hours?

3. A Toro lawn mower uses 5 tanks of gas to cut 18 acres of lawn. How many acres could be cut using 8 tanks of gas?

4. If 4 small pizzas cost $15.00, find the cost of 7 small pizzas.

5. In the first 4 games of the season, the Hurricanes Football team scored a total of 68 points. At this rate, how many points will the team score in 11 games?

6. A farmer knows that of every 50 eggs his chickens lay, only 45 will be sold. If his chickens lay 1000 eggs in a week, how many of them will be sold?

7. Alonzo Mourning scored 162 points in 9 games. At this rate, how many points will he score in 20 games.

8. At Rainbow TV Service, Al can repair 6 televisions in 2 weeks. How many televisions can Al repair in 10 weeks?

9. In a 10 game season, Willis McGahee rushed for 1250 yards. On the average how many yards did he rush for in 5 games?

10. If 8 bandanas cost $22.00, how much would 12 bandanas cost?

11. Adrienne paid $34.50 for 3 books. How much would she pay for 11 books?

12. American Idol auditioned 18,500 people in 2 cities. At that same rate, how many people auditioned in 7 cities?

13. Seven cups of oatmeal contain 378 grams of carbohydrates. How many grams of carbohydrates would there be in 4 cups of oatmeal?

14. If Skye uses 10 apples to make 3 apple pies, how many pies could Skye make if she uses 30 apples?

15. A male astronaut that weighs 195 pounds on Earth would weigh 33 pounds on the moon. If a female astronaut weighs 17 pound on the moon, what would her weight be on Earth? (Round your answer to the nearest pound.)

16. To make a really great punch, mix 2 quarts of Sprite with 6 quarts of pineapple sherbet. How many quarts of Sprite would you need when you use 11 quarts of pineapple sherbet?

Proportions   Word Problem Workbook

17. In the course of 2 days, Lifetime TV airs 5 reruns of *The Nanny*. How many reruns would air over the course of 12 days?

18. Amber, a standard poodle, can run around the block 5 times in 12 minutes. If she always runs at the same speed, how many times could she run around the block in 48 minutes?

19. If it takes Kelli 20 minutes to learn one dance routine, how long would it take her to learn 6 dance routines?

20. On a road map, 4 inches represents 50 miles. How many inches would represent 125 miles?

Section 4

Parts

Step 1
Read Through The Entire Problem

It is important to read the problem through in order to get a sense of the information given in the problem and what question is being asked.

You can usually recognize Parts Problems because they:

- State a scenario using **Direct Translation Words**.
- Often use the word "**sum**" or "**total**" in the problem.
- Give a **whole** amount which is **separated into different parts** or items.

Step 2
Name The Expressions Using Direct Translation & Use A Variable For The Totally Unknown

In Parts Problems, you will be given two or three items (or parts) that added together would make up a whole. For example, a certain amount of boys and a certain amount of girls added together make up the total amount of students in a class.

Although it may sometimes seem you do not have enough information to solve the problem, you will learn how to name a **Totally Unknown** and to use Direct Translation to name the other expressions to set up an equation.

For example, if the problem says "the class contained 5 more boys than girls", you are getting *some* information about the boys (there are 5 more boys than girls). You are not getting *any* information about the girls.

Since you are not getting any information about the girls, the girls become the **Totally Unknown**. Therefore, you will represent "the girls" with a variable, such as x.

Once you name the **Totally Unknown** with a variable, you can build upon that variable using Direct Translation in order to name the other expressions.

HELPFUL HINTS

- If you are not sure which part is the **Totally Unknown**, look for the word at the end of the phase or sentence *that gives you information about the parts*.

- In the above example, "the class contained 5 more boys than girls", the word "girls" is at the end of the parts information phrase or sentence. Therefore, you can confirm that "**girls**" is the **Totally Unknown**.

Parts   Word Problem Workbook

Step 3
Set Up An Equation

The total amount will always be given in a Parts Problem. So once you name the expressions for each part, you will have all the information you need. Set up the equation by adding all of the expressions together set equal to the total amount given.

Step 4
Solve the Equation

Using the method taught by your instructor, solve the equation for the variable.

Step 5
Make Sure to Answer the Question Being Asked

When you solve the equation, you will find the value for x, but that might not be the answer to the question. You need to re-read the problem and make sure exactly what question is being asked.

It is possible that the value for the variable x may be your answer. But it may *not* be.

For example, the value for x may give you the amount of girls in the class, while the problem may be asking for the amount of boys. You will have to substitute the value for x into the original expression for boys that you set up in Step 2 in order to get the correct answer. Always be sure of exactly what the question is.

HELPFUL HINT
- Once you have the numerical value for each part, you can always check if your answer is correct by adding up all the numerical values. Your total should equal the total of the whole amount given in the problem.

Word Problem Workbook   Parts

EXAMPLES

EXAMPLE 1 Jessica and Natalie are sisters. Jessica is 3 years younger than Natalie. If the sum of their ages is 25, how old is Jessica?

SOLUTION

Step 1 *Read The Problem*

- The problem tells a story and uses the word "sum".
- Jessica's age and Natalie's age are two parts of the whole (the sum of their ages).
- The answer wanted is Jessica's age.
- The problem contains Direct Translation Words [younger than, is].

Step 2 *Name The Expressions*

- You are given information about Jessica's age – she is 3 years younger than Natalie.
- You are not given information about Natalie's age. Her age is the Totally Unknown.
- Confirm Totally Unknown by checking position of word "Natalie" (Step 2 Helpful Hint).
- Use a variable to represent the Totally Unknown (in this example, Natalie's age).
- Use Direct Translation to name an expression for Jessica's age.

$$\text{Natalie's age} = x$$
$$\text{Jessica's age} = x - 3$$

Step 3 *Set Up The Equation*

- Natalie's age plus Jessica's age equals the total (the sum of their ages).
- Use the expressions you named in step 2 for their ages.
- The sum of their ages is 25. This amount is given in the problem.

$$x + x - 3 = 25$$

Step 4 *Solve The Equation*

- The solution to the equation is $\boxed{x = 14}$

Step 5 *Answer The Question Asked*

- You have the solution to the equation, but it is NOT the answer to the question.
- The value of *x* is Natalie's age; the problem asks "How old is *Jessica?*"
- You need to use the expression for Jessica's age that you named in Step 2.
- Get the answer by substituting the solution for *x* (which is 14) into that expression.

$$\begin{aligned} \text{Jessica's age} &= x - 3 \\ \text{Jessica's age} &= 14 - 3 \\ \text{Jessica's age} &= 11 \end{aligned}$$

Answer: Jessica is 11 years old.

HELPFUL HINT

- Sometimes Direction Translation Words are separated by other words such as "seven **less** Sophomores **than** Juniors" (as in the next example). These are still Direction Translation Words and should be translated as though they were not separated.

Word Problem Workbook   Parts

EXAMPLE 2   The Boyd Anderson High School Band has a total of 93 members. The band has 7 less Sophomores than Juniors and 10 more Seniors than Juniors. How many members of the Boyd Anderson High School Band are Seniors?

SOLUTION

Step 1 *Read The Problem*

- The problem tells a story and uses the word "total".
- The Sophomores, Juniors, and Seniors are 3 parts of the whole (the total band members).
- The answer wanted is the amount of members that are Seniors.
- The problem contains Direct Translation Words [less than, more than].

Step 2 *Name The Expressions*

- You have information about the Sophomores; there are 7 less Sophomores than Juniors.
- You have information about the Seniors; there are 10 more Seniors than Juniors.
- You have no information about the Juniors, so that amount is the Totally Unknown.
- Confirm Totally Unknown by checking position of word "Juniors" (Step 2 Helpful Hint).
- Use a variable to represent the Totally Unknown (in this example, the Juniors).
- Use Direct Translation to name the two expressions for the Sophomores and Seniors.

$$\text{Juniors} = x$$
$$\text{Sophomores} = x - 7$$
$$\text{Seniors} = x + 10$$

Step 3 *Set Up The Equation*

- The Juniors plus the Sophomores plus the Seniors equal the total (the sum of members).
- Use the expressions you named in Step 2 for the different members of the band.
- The sum of all the members in the band is 93. This amount is given in the problem.

$$x + x - 7 + x + 10 = 93$$

Step 4 *Solve The Equation*

- The solution to the equation is $\boxed{x = 30}$

Step 5 *Answer The Question Asked*

- You have the solution to the equation, but it is NOT the answer to the question.
- x is the amount of Juniors; the question asks how many band members "are Seniors?"
- You need to use the expression for Seniors that you named in Step 2.
- Get the answer by substituting the solution for x (which is 30) into that expression.

$$\boxed{\begin{array}{l} \text{Seniors} = x + 10 \\ \text{Seniors} = 30 + 10 \\ \text{Seniors} = 40 \end{array}}$$

Answer: There are 40 Seniors in the band.

Section 4: Parts Exercise Set

1. Jacob is 7 years older than twice Sarah's age. If the sum of their ages is 61, how old is Jacob?

2. On an Algebra test, the highest grade was 38 points higher than the lowest grade. The sum of the two grades was 142. Find the lowest grade.

3. A History textbook costs $7 less than a Sociology textbook. If the total cost of both textbooks is $73, what is the cost of the history textbook?

4. A Pre-Algebra class contains a total of 45 students. If the number of women is 3 less than twice the number of men, how many women and how many men are in the class?

5. A box of candy contains 28 pieces. If the number of pieces of milk chocolate is 4 less than 3 times the number of pieces of dark chocolate, how many pieces of each kind are there?

6. Victor Gil runs a ski train. One day he noticed that the train contained 13 more women than men. If there were a total of 165 people on the train, how many of them were women?

7. On a Geometry test, the highest grade was 57 points higher than the lowest grade. The sum of the two grades was 95. Find the highest and lowest grades.

8. Alexandra and Valery sold a total of 348 boxes of Girl Scout cookies. Alexandra sold 52 more boxes than Valery. How many boxes did Valery sell?

9. On a small cruise ship, there were 110 more women than men. If there were 910 people on the cruise ship, how many women were there?

10. In the 1992 Presidential Election, George Bush received 202 less electoral votes than his opponent, Bill Clinton. The electoral college has a total of 538 votes to cast. How many electoral votes did the winner, Bill Clinton, receive in the 1992 election?

11. At Central Park School Field Day, Jarred Suede threw a baseball 12 feet further than his friend, Brendan. If the sum of the distances of both of their throws was 52 feet, how far did Jarred Suede throw the baseball?

12. Lee spent $32 more on a CD player than she did on a scientific calculator. If the total amount that Lee spent was $62, how much did each item cost?

13. In the 2000 Senate race for Connecticut, the winner was Senator Joe Lieberman. He received 379,034 more votes than his opponent Phil Giordano. If a total of 1,269,658 votes were cast, how many votes did Senator Lieberman receive?

14. Antonio and Cristian swim at a local pool. Cristian usually swims 4 less than twice the amount of laps Antonio swims. If they swam a total of 44 laps altogether, how many laps did Cristian swim?

15. A 61-foot board will be cut into three pieces. The second piece will be 6 feet shorter than twice the first piece, and the third piece will be 7 feet longer than the first piece. How long will the second piece of the board be?

16. Irene paid $40 more for her cell phone bill in July than in June. In August, she paid $10 less than her June bill. If she paid a total of $120 for her cell phone bills for all three months, how much did she pay for her August bill?

17. A piece of string 37 inches long was cut into three pieces. The first piece is 7 inches shorter than the second piece, and the third piece is twice as long as the second piece. How long is each piece of the string?

18. Elena checked out a total of 20 books from the library. She checked out 4 more mystery novels than romance novels, and she checked out 5 less biographies than romance novels. How many of each type of book did Elena check out of the library?

19. Lafayette County in Florida has 4269 registered voters. There are 3428 more Democrats than 5 times the amount of Independents, and there are 497 more Republicans than twice the amount of Independents. How many of the voters in Lafayette county are registered Democrats?

20. In an episode of the Powerpuff Girls, Buttercup defeated twice as many villains as Blossom. Bubbles defeated 6 less villains than Blossom. If the Powerpuff Girls defeated a total of 74 villains, how many villains did Buttercup defeat?

Section 5

Geometric Perimeter

You can recognize a Geometric Perimeter problem because it uses the word "**perimeter**". Once you identify a Geometric Perimeter problem, use the method taught in this section to set up and solve the problem.

Step 1
Read Through The Entire Problem

Use this opportunity to identify the **geometric shape** in the problem (for example, triangle or rectangle), and determine **exactly what question is being asked**. Get a sense of the problem, and look for Direct Translation Words.

Step 2
Name The Expressions Using Direct Translation & Use A Variable For The Totally Unknown

In Perimeter Problems, you will be reading about the **length** and **width** and **sides** of geometric shapes. Use a variable for the Totally Unknown, and focus on the Direct Translation Words to build on the variable and name all of the remaining expressions.

For example, if the problem says "the length of a rectangle is 4 feet more than its width", it is telling you that the **length** is *something*: It is 4 feet more than its width.

You are getting some information about the length, but you are not getting any information about the width. This is how you know that the **width** is the **Totally Unknown**, so you will represent the **width** with a variable, such as x.

Step 3
Set Up An Equation By Substituting The Expressions Into The Appropriate Formula

The value for the perimeter is always given. So now that you can name the expressions for the sides of the shape, you have all the information that you need to substitute into the formula and set up the equation.

Geometric Shape	Formula
Rectangle	$2L + 2W = P$
Triangle	$a + b + c = P$

Geometric Perimeter   *Word Problem Workbook*

HELPFUL HINT

- If you can't recall the formula for finding the perimeter of a geometric shape, remember that "perimeter" means the distance around the outside of the shape. You can always set up your equation by adding together all of the sides of the shape and then setting it equal to the value of the perimeter.

Shape	Common Sense Formula To Use If You Can't Remember The Geometric Formula
Rectangle	Long Side + Long Side + Short Side + Short Side = Perimeter
Triangle	First Side + Second Side + Third Side = Perimeter

Step 4
Solve the Equation

Using the method taught by your instructor, solve the equation for the variable.

Step 5
Make Sure to Answer the Question Being Asked

As in Parts Problems in Section 4, when you solve the equation you will find the value for x, but that might not be the answer to the question. You need to reread the problem and make sure exactly what question is being asked.

It is possible that the value for the variable x may be your answer. But it may *not* be.

For example, the value for x may give you the *width* of a rectangle, while the problem may be asking for the *length* of the rectangle. You will have to substitute the value for x into the original expression for the length that you will set up in Step 2 in order to get the correct answer. Always be sure of exactly what the question is.

HELPFUL HINT

- Once you have the numerical value for each side, you can always check if your answer is correct by adding up all the sides of the shape. Your total amount should equal the perimeter given in the problem.

Word Problem Workbook   Geometric Perimeter

EXAMPLES

EXAMPLE 1 The length of a rectangle is 3 feet less than twice its width. If the perimeter of the rectangle is 30 feet, find its length.

SOLUTION

Step 1 *Read The Problem*

- Seeing the word "perimeter" verifies that it is a Geometric Perimeter problem.
- The geometric shape is a rectangle.
- The answer that is wanted is the length of the rectangle.
- The problem contains Direct Translation Words [less than, twice, is].

Step 2 *Name The Expressions*

- You are given information about the length; the length is 3 feet less than twice the width.
- You are not given information about the width, so the width is the Totally Unknown.
- Use a variable to represent the Totally Unknown (in this example, the width).
- Use Direct Translation to name an expression for the length.

$$\text{Width } (W) = x$$
$$\text{Length } (L) = 2x - 3$$

Step 3 *Set Up The Equation*

- Use the formula for a rectangle, $2L + 2W = P$.
- The perimeter is 30, so substitute the number 30 for P.
- Substitute the expressions you named in Step 2 for L and W into the formula.

$$2(2x - 3) + 2(x) = 30$$

Geometric Perimeter   Word Problem Workbook

Step 4 *Solve The Equation*

- The solution to the equation is $\boxed{x = 6}$

Step 5 *Answer The Question Asked*

- You have the solution to the equation, but it is NOT the answer to the question.
- The value of *x* is the size of the width; the problem states "find its *length*".
- You need to use the expression for the length that you named in Step 2.
- Get the answer by substituting the solution for *x* (which is 6) into the expression.

$$\boxed{\begin{aligned} L &= 2x - 3 \\ L &= 2(6) - 3 \\ L &= 9 \end{aligned}}$$ *Answer*: The length of the rectangle is 9 feet.

Word Problem Workbook   Geometric Perimeter

EXAMPLE 2   On the show Trading Places, designer Vern Yip constructs a large triangular mirror for a living room. The length of the longest side is twice the length of the middle side. The shortest side is 4 feet shorter than the middle side. If the perimeter of the mirror is 24 feet, what is the length of the longest side?

SOLUTION

Step 1 *Read The Problem*

- Seeing the word "perimeter" verifies that it is a Geometric Perimeter problem.
- The geometric shape is a triangle.
- The answer that is wanted is the length of the longest side.
- The problem contains Direct Translation Words [twice, is, shorter than].

Step 2 *Name The Expressions*

- You have information about the longest side; it is twice the length of the middle side.
- You have information about the shortest side; it is 4 feet shorter than the middle side.
- You have no information about the middle side, so it is the Totally Unknown.
- Use a variable to represent the Totally Unknown (in this example, the middle side).
- Use Direct Translation to name the expression for the longest side and the shortest side.

$$\text{Middle Side } (a) = x$$
$$\text{Longest Side } (b) = 2x$$
$$\text{Shortest Side } (c) = x - 4$$

Step 3 *Set Up The Equation*

- Use the formula for a triangle, $a + b + c = P$
- The perimeter is 24, so substitute the number 24 for P.
- Substitute the expressions you named in Step 2 for $a, b,$ and c into the formula.

$$x + 2x + x - 4 = 24$$

Geometric Perimeter

Step 4 *Solve The Equation*

- The solution to the equation is $x = 7$

Step 5 *Answer The Question Asked*

- You have the solution to the equation, but it is NOT the answer to the question.
- x is the size of the middle side; the question asks "what is the length of the longest side?"
- You need to use the expression for the longest side (b) that you named in Step 2.
- Get the answer by substituting the solution for x (which is 7) into the expression.

$$b = 2x$$
$$b = 2(7)$$
$$b = 14$$

Answer: The longest side of the triangle is 14 feet.

Section 5: Geometric Perimeter Exercise Set

1. The perimeter of a rectangle is 40 meters. The width is 4 meters less than the length. What is the length of the rectangle?

2. The length of a rectangle is 11 feet less than twice its width. Its perimeter is 26 feet. What is the length and width of the rectangle?

3. The length of a rectangle is 7 inches less than 3 times its width. The perimeter of the rectangle is 50 inches. Find the length of the rectangle.

4. The length of a rectangular parking lot is 10 meters less than twice its width, and the perimeter is 400 meters. Find the length of the parking lot.

5. The first side of a triangle is 3 centimeters longer than the second side. The third side of the triangle is 4 centimeters shorter than the second side. If the perimeter of the triangle is 32, how long is the first side?

6. In a triangle, the length of the second side is twice the length of the first side. The length of the third side is 8 less than 3 times the length of the first side. If the perimeter of this triangle is 34 inches, how long is each side?

7. The shortest side of a triangle is 5 yards shorter than the longest side. The middle side is 3 yards shorter than the longest side. If the perimeter of the triangle is 28 yards, find the length of the shortest side.

8. The third side of a triangle is one foot less than 4 times the length of the first side of the triangle. The second side is 6 feet more than the first side. If the perimeter of the triangle is 77 feet, what is the length of all three sides?

9. The width of a rectangle is 7 feet less than its length. If the perimeter of a rectangle is 58 feet, find the length and width of the rectangle.

10. The length of a rectangular yard is 12 feet more than twice its width. If the perimeter of the yard is 276 feet, find the width.

11. The largest known map is a relief map of Asia. It is rectangular in shape and its length is 25 times its width. The perimeter is 936 feet. What is the length of the map?

12. The length of a rectangle is 20 inches more than its width. If the perimeter is 176 inches, find the width of the rectangle.

13. The longest side of a triangle is 3 times the length of the shortest side. The middle side is 3 feet longer than the shortest side. If the perimeter of the triangle is 53 feet, what is the length of the middle side?

Geometric Perimeter Word Problem Workbook

14. The middle side of a triangle is 7 inches longer than twice the shortest side. The longest side is 3 inches more than 5 times the shortest side. If the perimeter of the triangle is 82 inches, find the length of all three sides.

15. Sierra painted a ceramic triangle to hang on her bedroom wall. Two sides of the triangle are of equal length, and the third side is 6 inches longer than the length of one of the other sides. The perimeter of her ceramic triangle is 33 inches. What is the length of the third side? (Hint: If two sides are equal, use the same variable for each one.)

16. The shortest side of a triangle is 6 inches less than the middle side. The longest side is 11 inches less than 5 times the middle side. What is the length of the longest side of the triangle if the perimeter is 53 inches?

17. The length of a rectangle is 10 centimeters greater than its width. Find the length and width of the rectangle if its perimeter is 168 centimeters.

18. In the entranceway of the Sunrise Cinema, there is a large rectangular poster advertising the upcoming movie "Terminator III". If the poster's length is 3 feet less than twice its width, and its perimeter is 24 feet, what is the length of the poster?

19. The Chesterbrook Academy playground is rectangular in shape. Its length is 4 times its width. If the perimeter of the playground is 70 feet, what is its length and width?

20. Jarred's little league soccer team plays on a rectangular field with a perimeter of 78 yards. If the length of the field is 15 yards longer than its width, what are the dimensions of the soccer field?

Section 6

Consecutive Integers

You will always see the word "**consecutive**" in a Consecutive Integers problem. Consecutive means "directly following", so a Consecutive Integer means the next whole number. For example, if you start with the number "6", the next Consecutive Integer would be "7".

Step 1
Read Through The Entire Problem

It is extremely important to carefully read through a Consecutive Integers Problem, because there are two important facts you need to determine before moving on to the next step.
- Will you be using consecutive integers, consecutive *odd* integers, or consecutive *even* integers?
- How many integers will you be using in the equation?

Step 2
Name The Expressions For The Integers

There are two ways to set up the expressions for consecutive integers. This is determined by whether the problem is about plain consecutive integers, or whether it is about either consecutive *odd* or *even* integers.

The setup will begin with an unknown, expressed by the variable x. That variable *always* represents the first integer. In a problem dealing with plain consecutive integers, the second integer would be the number right after x, which would make it $x + 1$. The third integer would be one more than the second, making it $x + 1 + 1$, or $x + 2$.

In problems dealing with consecutive odd or even integers, the variable x still represents the first integer. But because we are working with either odd or even numbers, we are skipping over a number to get to the next desired integer.

Therefore, in a problem dealing with consecutive odd or even integers, the second integer would be 2 numbers after the x, which would make it $x + 2$. The third integer would be two more than the second integer, making it $x + 2 + 2$ or $x + 4$.

Helpful Hint

- In all Consecutive Integer Problems, the **first integer** may also be referred to as the **smaller** or **smallest**.
- The **second integer** is referred to as the **middle**, except if there are only two integers in the problem. *(Then the second integer would be the larger or largest)*.
- The **third integer** will also be referred to as the **larger** or **largest**.

Consecutive Integers   *Word Problem Workbook*

Step 3
Set Up An Equation

With Consecutive Integers Problems, there are two different ways to set up the equation.

- If the word "sum" is stated in the problem, add up the expressions for the integers you set up in Step 2 and set them equal to the given sum.

- If you don't see the word "sum", you will see other Direct Translation Words. In that case, you use Direct Translation in order to set up an equation. (Review Section 1 if necessary.)

Step 4
Solve the Equation

Using the method taught by your instructor, solve the equation for the variable.

Step 5
Make Sure to Answer the Question Being Asked

As in previous Chapters, when you solve the equation you will find the value for x, but that might not be the answer to the question. You need to reread the problem and make sure exactly what question is being asked.

It is possible that the value for the variable x may be your answer. But it may *not* be.

For example, the value for x will always give you the *first* integer, while the problem may be asking for the *second* or the *largest* integer. In order to get the correct answer, you will have to substitute the value for x into the original expression for the desired integer that you will set up in Step 2. Always be sure of exactly what the question is.

Helpful Hint

- You may find the common sense approach easier in order to determine the correct answer. For example, if $x = 7$, and the question is asking for the next odd integer, you know that "9" is the second (middle) odd integer, and "11" is the third (largest).

CONSECUTIVE INTEGER TABLE

Integer	Setup for Plain Consecutive Integers	Setup for Odd or Even Consecutive Integers	EXAMPLE 1st Integer is 5 (Plain Integers)	EXAMPLE 1st Integer is 11 (odd integers)	EXAMPLE if 1st Integer is 8 (even integers)
First	x	x	5	11	8
Second	$x + 1$	$x + 2$	$5 + 1 = 6$	$11 + 2 = 13$	$8 + 2 = 10$
Third	$x + 2$	$x + 4$	$5 + 2 = 7$	$11 + 4 = 15$	$8 + 4 = 12$

EXAMPLES

EXAMPLE 1 The sum of three consecutive integers is 72. Find the three integers.

SOLUTION

Step 1 *Read The Problem*

- Seeing the word "consecutive" verifies that it is a *plain* Consecutive Integers problem.
- The problem states "consecutive integers". There is no mention of Odd or Even.
- The problem states that there are *three* consecutive integers.

Step 2 *Name The Expressions*

- The first integer is x.
- These are plain consecutive integers, so the second integer is $x + 1$.
- The third integer is $x + 2$.

$$\begin{aligned} \text{First} &= x \\ \text{Second} &= x + 1 \\ \text{Third} &= x + 2 \end{aligned}$$

Consecutive Integers

Step 3 Set Up The Equation

- The word "sum" is used in the example.
- Add the expressions you named for the integers in step 2.
- Set the expressions equal to the total (72) which is given.

$$x + x + 1 + x + 2 = 72$$

Step 4 Solve The Equation

- The solution to the equation is $x = 23$

Step 5 Answer The Question Asked

- You have the solution to the equation, but it is NOT the answer to the question.
- The value of x is the 1st integer; but you need to find all three integers.
- You need to use the expressions for the 2nd and 3rd integers that you named in Step 2.
- Get the answer by substituting the solution for x (which is 23) into the expressions.
- You may use common sense to get the 2nd and 3rd integers (Step 5 Helpful Hint).

2nd integer $= x + 1$	3rd integer $= x + 2$
2nd integer $= 23 + 1$	3rd integer $= 23 + 2$
2nd integer $= 24$	3rd integer $= 25$

Answer: The 3 integers are 23, 24, 25

Word Problem Workbook — *Consecutive Integers*

EXAMPLE 2 The sum of three consecutive odd integers is 63. Find the largest of the three integers.

SOLUTION

Step 1 *Read The Problem*

- Seeing the word "consecutive" verifies that it is a Consecutive Integers problem.
- The problem states "consecutive *odd* integers".
- The problem states that there are three consecutive integers.

Step 2 *Name The Expressions*

- The first integer is x.
- These are consecutive odd integers, so the second integer is $x + 2$.
- The third integer is $x + 4$.

$$\begin{aligned} \text{First} &= x \\ \text{Second} &= x + 2 \\ \text{Third} &= x + 4 \end{aligned}$$

Step 3 *Set Up The Equation*

- The word "sum" is used in the example.
- Add the expressions you named for the integers in step 2.
- Set the expressions equal to the total (63) which is given.

$$x + x + 2 + x + 4 = 63$$

Consecutive Integers — Word Problem Workbook

Step 4 *Solve The Equation*

- The solution to the equation is $x = 19$

Step 5 *Answer The Question Asked*

- You have the solution to the equation, but it is NOT the answer to the question.
- The value of x is the 1st (smallest) integer, but you need to find the largest integer.
- You need to use the expressions for the largest (3rd) integer that you named in Step 2.
- Get the answer by substituting the solution for x (which is 19) into the expression.
- You may use common sense to get the largest integer (Step 5 Helpful Hint).

$$\text{Largest Integer} = x + 4$$
$$\text{Largest Integer} = 19 + 4$$
$$\text{Largest Integer} = 23$$

Answer: The largest integer is 23.

EXAMPLE 3   Find two consecutive integers such that the larger is 8 less than twice the smaller.

SOLUTION

Step 1 *Read The Problem*

- Seeing the word "consecutive" verifies that it is a Consecutive Integers problem.
- The problem states "consecutive *even* integers".
- The problem states that there are *two* consecutive integers.
- There are Translation Words [less than, is, twice].

Step 2 *Name The Expressions*

- The first (smaller) integer is x.
- There are only two integers so the second integer is the larger integer.
- These are consecutive even integers so the second (larger) integer is $x + 2$.

$$\text{First (smaller)} = x$$
$$\text{Second (larger)} = x + 2$$

Word Problem Workbook — *Consecutive Integers*

Step 3 *Set Up The Equation*

- You do NOT see the world "sum". You DO have Direct Translation Words.
- Use the expressions you named for the integers in Step 2.
- Set up an equation using Direct Translation. (Review Section 1 if necessary.)
- The larger integer will be set equal to 8 less than twice the smaller.

$$x + 2 = 2x - 8$$

Step 4 *Solve The Equation*

- The solution to the equation is $x = 10$

Step 5 *Answer The Question Asked*

- You have the solution to the equation, but it is NOT the answer to the question.
- The value of x is the 1st (smallest) integer, but you need to find both integers.
- You need to use the expressions for the second (larger) integer that you named in Step 2.
- Get the answer by substituting the solution for x (which is 10) into the expression.
- You may use the common sense approach to get the larger integer (Step 5 Helpful Hint).

Larger integer $= x + 2$
Larger integer $= 10 + 2$
Larger integer $= 12$

Answer: The larger integer is 12.

Section 6: Consecutive Integers Exercise Set

1. The sum of two consecutive integers is 41. Find the integers.
2. The sum of two consecutive integers is 193. Find the integers.
3. The sum of two consecutive odd integers is 32. Find the integers.
4. The sum of two consecutive odd integers is 104. Find the integers.
5. The sum of two consecutive even integers is 66. Find the integers.
6. The sum of two consecutive even integers is 90. Find the integers.
7. The sum of three consecutive integers is 93. Find the integers.
8. The sum of three consecutive integers is 120. Find the integers.
9. The sum of three consecutive odd integers is 75. Find the integers.
10. The sum of three consecutive odd integers is 153. Find the integers.
11. The sum of three consecutive even integers is 72. Find the integers.
12. The sum of three consecutive even integers is 138. Find the integers.
13. Find the second of two consecutive integers if the second is 13 less than twice the first.
14. Find the larger of two consecutive integers if the larger is 149 less than twice the smaller.
15. Find the smaller of two consecutive odd integers if the larger is 20 less than three times the smaller.
16. Find the larger of two consecutive odd integers if the larger is 40 less than three times the smaller.
17. Find the larger of two consecutive even integers if twice the smaller is 48 more than the larger.
18. Find the smaller of two consecutive even integers if three times the smaller is 8 more than twice the larger.
19. Find the second of three consecutive integers if the sum of twice the first and 4 times the second is equal to 20 more than twice the third.
20. Find three consecutive integers if the third is equal to 15 less than the sum of the first and second.

Section 7

Motion – Linear Equations

For some students, these are the problems that give them the worst headaches. You have trains, planes, buses, joggers, or anything that moves, going in the same direction, going in opposite directions – and then you're asked questions like:

- How long will it take before they meet?
- How long will it be before they're 200 miles apart?
- How fast were they going?

When you finish this section, you will learn how to identify the important bits of information and what to do with it so that you will be able to make sense of these problems and solve them.

Step 1
Read Through The Entire Problem

In these Motion Problems, you have two objects in motion traveling in the same or opposite directions. You need to look for and take note of any information that refers to:
- The **time of travel**
- The **speed of the objects**
- The **miles apart the objects are**
- The **direction** the objects are moving (the same or opposite).

Step 2
Fill In The Chart Using The Distance Formula

Motion problems rely heavily on the distance formula, "**distance (d) = rate (r) • time (t)**".

Drawing a pre-equation chart like the one below will organize your information so that you can set up your equation. The chart always has two rows which represents the two objects in motion.

r	•	t	=	d

Enter in your chart the information in the problem that refers to **rate** (speed) and **time**.

The **rate** in the problem is given in one of two ways. It is either given as a numerical speed (for example, 35 mph) or you will need to use Direct Translation to name an expression for each rate.

Motion – Linear Equations Word Problem Workbook

In this section, the **amount of time for both moving objects** will always be the *same*.

- If the amount of time is given, put that amount in the Time Column (*t*) for both rows.

- If the amount of time is not given, use the same variable for both.

Distance is not given in the word problem. You will get the expression to fill in the distance portion of the chart by using the distance formula and multiplying together the rate and time you enter.

HELPFUL HINT
- Do not confuse **Distance** with **Miles Apart**. Although "Miles Apart" represents a distance between two objects and IS given in the word problem, the "distance" in the formula represents how far each of the object has traveled. You will be using the amount of "Miles Apart" in the next step to set up the equation.

Step 3
Set Up Tthe Equation Based On The Information In The Chart

Now that you have created your chart, you have a snapshot of the word problem and the important information you need to set up the equation. It is important to note that **not everything from your chart goes into the equation**.

To determine the equation, you will ned the two distances from the chart and the "miles apart" that you are given in the problem.

For the Motion Problems in this section, there are only **two possible ways to set up an equation**.

If the two objects are traveling in **opposite directions**, *add the two distances from the chart* set equal to the miles apart.

If the two objects are traveling in the **same direction**, *subtract the smaller distance from the larger* set equal to the miles apart.

Direction	Equation
Opposite	distance + distance = miles apart
Same	distance − distance = miles apart

> **HELPFUL HINT**
>
> - It is important to be aware that opposite does not necessarily mean traveling away from each other. Even though objects are going toward each other, *they are still moving in opposite directions.*

Step 4
Solve the Equation

Using the method taught by your instructor, solve the equation for the variable.

Step 5
Make Sure to Answer the Question Being Asked

In Motion Problems, as in other word problems, you need to make sure exactly what question is being asked. It is possible that the value for the variable x may be your answer. But it may *not* be. For example, the value for x may be the time, or it may be one of the two rates. If the question asks for "time", you are done. The variable x will be the answer. If the question asks for "rate", you need to see which rate is being asked for in the problem, and substitute the value of x in that rate in the chart.

Motion – Linear Equations

Word Problem Workbook

EXAMPLES

EXAMPLE 1 Two steamboats leave a city at the same time traveling in the same direction on a straight canal. One travels at 18 miles per hour and the other travels at 25 miles per hour. In how many hours will the boats be 35 miles apart?

SOLUTION

Step 1 *Read The Problem*

- There are two objects in motion, traveling; it is a Motion – Linear Equation Problem.
- The rate (speed) of each object is given. The miles apart is given.
- The time is *not* given; that will be your variable.
- The two objects are traveling in the *same* direction.

Step 2 *Fill In The Chart*

- In your chart, fill in the rates given for each steamboat. One rate is 18 and one is 25.
- In your chart, fill in the time for both steamboats with the variable x.
- In your chart, fill in the distance for each steamboat by multiplying its rate and time.

Object	r	$\cdot$	t	$=$	d
Steamboat 1	18		x		$18x$
Steamboat 2	25		x		$25x$

Step 3 *Set Up The Equation*

- The problem states the steamboats are moving in the *same* direction.
- Use the expressions for the two distances that you named in your chart in Step 2.
- Miles Apart is given in the problem. It is 35.
- Set up your equation as distance – distance = miles apart.

$$25x - 18x = 35$$

Word Problem Workbook   Motion – Linear Equations

Step 4 *Solve The Equation*

- The solution to the equation is $\boxed{x = 5}$

Step 5 *Answer The Question Asked*

- You have the solution to the equation. The value of x is 5.
- The question asked is "how many hours?" The question is asking for the time.
- When the question asks for "time", you are done. The value of x is the answer.

    Answer: In 5 hours

EXAMPLE 2   Two trains leave the station at the same time. One train is traveling North at 10 mph faster than the other train which is traveling South. After 6 hours, the two trains are 720 miles apart. At what speed did the faster train travel?

SOLUTION

Step 1 *Read The Problem*

- There are two objects in motion, traveling; it is a Motion – Linear Equation Problem.
- The time is given. The miles apart is given.
- The rates are *not* given. You will name the expressions for the rates.
- The two objects are traveling in *opposite* directions.

Step 2 *Fill In The Chart*

- In your chart, fill in the same time given for each train. It is 6 hours.
- The rate of the slower train is the Totally Unknown. Use the variable x.
- Use Direct Translation to name the expression for the rate of faster train. It is $x + 10$.
- In your chart, fill in the distance for each train by multiplying its rate and time.

Object	r	$\cdot$	t	$=$	d
Slower Train	x		6		$6x$
Faster Train	$x + 10$		6		$6(x + 10)$

Motion – Linear Equations   Word Problem Workbook

Step 3 *Set Up The Equation*

- The problem states the trains are moving in *opposite* directions.
- Use the expressions for the two distances that you named in your chart in Step 2.
- Miles Apart is given in the problem. It is 720.
- Set up your equation as distance + distance = miles apart.

$$6x + 6(x+10) = 720$$

Step 4 *Solve The Equation*

- The solution to the equation is $x = 55$

Step 5 *Answer The Question Asked*

- You have the solution to the equation. But that is NOT the answer to the problem.
- The value of x (55) is for the rate of the slower train.
- The question asked for is "the rate of the faster train".
- Substitute 55 for x into the expression in your chart for the rate of the faster train.

Faster train $= x + 10$
Faster train $= 55 + 10$
Faster train $= 65$

Answer: The rate is 65 mph

Section 7: Motion – Linear Equations Exercise Set

1. Beth and Jade begin rollerblading at the same time going in opposite directions. Beth is traveling at 10 mph, and Jade is traveling at 6 mph. How long will it be before they are 32 miles apart?

2. Two planes leave Miami International Airport at the same time flying in opposite directions. One plane is traveling at a speed of 550 mph and the other plane is traveling at a speed of 600 mph. In how many hours will they be 5750 miles apart?

3. Two thoroughbred horses begin a marathon at the same time, heading in the same direction. One horse gallops at a speed of 7 mph while the other horse gallops at a speed of 12 mph. How long will it be before the two greyhounds are 10 miles apart?

4. On University Drive in Lauderhill, two buses leave the bus stop at the same time traveling in the same direction. One bus is traveling at 40 mph and the other bus is traveling at 45 mph. How long will it be before they are 30 miles apart?

5. Two trains leave a city at the same time. One travels north at 60 mph and the other travels south at 80 mph. In how many hours will they be 420 miles apart?

6. Sharon and Bentley leave BCC at the same time traveling in cars going in opposite directions. Sharon travels at 40 mph and Bentley travels at 60 mph. In how many hours will they be 350 miles apart?

7. Trent and Evan are 72 miles apart. Both start riding their bicycles at the same time and travel toward each other. If Trent bikes at 13 mph and Evan bikes at 11 mph, in how many hours will they meet?

8. Two cross country skiers are 45 miles apart at opposite ends of a course. The two skiers start at the same time and travel towards each other. One skier is traveling at 14 mph and the second skier is traveling at 16 mph. How long will it take them before they meet?

9. John Elway left Denver on a plane flying at a speed of 350 mph traveling northwest to Seattle. His coach, Mike Shanahan, left Denver at the same time on a plane flying at a speed of 420 mph traveling southeast to Atlanta. How many hours did it take until their flights were 1925 miles apart?

10. Two high school friends leave Fort Lauderdale at the same time to go to college. One is driving North to Gainesville at a speed of 65 mph. Her friend is driving South to the University of Miami at a speed of 55 mph. How long will it be before the two friends are 180 miles apart?

11. Two joggers are 10 miles apart. At the same time, then begin running towards each other. One jogger is running at 4 mph and the other is running at 6 mph. How long after they begin will they meet?

Motion – Linear Equations   *Word Problem Workbook*

12. An airplane leaves Kennedy Airport in New York headed for Los Angeles International Airport in California. The plane is flying at a speed of 450 mph. Another airplane leaves Los Angeles International Airport headed for Kennedy Airport in New York, flying at a speed of 500 mph. If the two airports are 2375 miles apart, when will the two planes pass each other in flight?

13. The band members of Walton High School in the Bronx are going to a band competition in Florida. They travel on two buses that leave at the same time going in the same direction. One bus is traveling at a speed of 55 mph and the other is driving at a speed of 45 mph. How long will it take for the buses to be 50 miles apart?

14. A freight train and a passenger train both leave a train station at the same time, traveling in the same direction. The speed of the freight train is 45 mph and the speed of the passenger train is 65 mph. How long will it take before the two trains are 70 miles apart?

15. Victor and Ingrid leave their house at the same time riding their bicycles in opposite directions. Victor rides 5 mph faster than Ingrid. After two hours, they are 58 miles apart. At what speed was Victor riding?

16. Frances and Larry live 306 miles apart. They start driving towards each other and meet in three hours. If Frances drives 12 mph faster than Larry, find Larry's speed.

17. Two hot air balloons are 140 miles apart and are traveling towards each other. One hot air balloon is traveling 15 mph faster than the other. After 4 hours of traveling, they meet. What was the speed of each hot air balloon?

18. Vivian and her sister, Vicky, start jogging at the same time, going in opposite directions. Vivian jogs 3 mph faster than Vicky. After jogging for 3 hours, they are 39 miles apart. At what speed were both sisters jogging?

19. Randi and her daughter, Sierra, are white water rafting in Colorado. At the start of the day they are 160 miles apart. They begin traveling toward each other at the same time. Sierra's raft is traveling 8 mph faster than Randi's. If they meet in 5 hours, how fast was Randi's raft traveling?

20. Two cross country skiers begin at the same time, traveling in opposite directions. One skier is traveling at a rate of 6 mph slower than the other skier. In 8 hours, they are 96 miles apart. At what speed was the faster skier traveling?

Section 8

Money

Money problems involve different denominations. How many five-dollar bills are in your wallet? Do you have more quarters than nickels? In this section, you will learn to solve these problems.

8.1 Definition Of A Denomination Of Money

When the word "Denomination" is used in connection with money, it is **defining different numerical values** of money. Money includes both bills and coins.

With coins, you have a different value according to each type of coin. Therefore, each different type is a different denomination. A penny is one denomination, a nickel is another, dimes and quarters are two more. If someone has three denominations of coins in their hand, you know they are holding three different kinds of coins.

The same holds true for paper money (bills). Each different value is a different denomination of money. A one-dollar bill is one kind of denomination. A five-dollar bill is another kind of denomination, and so on.

If you only had twenties in your wallet, you would say you have only one denomination. All of your bills would be the same denomination. If you only had dimes in your hand, you would say the same -- you only have one denomination of money in your hand, because all your coins would be the same value.

To summarize, each kind of bill or coin that has a different value is called a denomination of money.

8.2 Value Of Denominations

Each denomination, whether bills or coins, has a value that you use in Money Word Problems.

- With coins, you use the decimal value of a single coin.
- With bills you use the whole number value of a single bill.

For example, if the problem states "fives" or "five-dollar bills", you would normally translate this as $5.00. However, since the decimal places are always zeros, you do not need the decimal point at all.

To simplify matters, just use the whole number "5". If the problem involved "twenties" or "twenty-dollar bills", use the whole number "20". The same would hold true for any bill denomination.

DENOMINATION VALUE CHART

If Problem States	Value Of Denomination (A Single Bill Or Coin)
Pennies	.01
Nickels	.05
Dimes	.10
Quarters	.25
Bills	Use Its Whole Number

8.3 Defining The Total Number Of Each Denomination & The Total Worth

With Money Problems, you have to be careful not to confuse the **number** of each denomination (bills or coins) with the **worth** of the money. In Money Word Problems, the **Total Worth** is given to you.

- *Number* has to do with "**how many**".
- *Worth* has to do with "**how much**".

When you discuss the **number** of a denomination, this is referring to "**how many**" you have of a particular denomination, NOT how much it is worth.

For example, if you have six quarters and three dimes, you have **two** denominations:

- The **value** of the 1st denomination (quarters) is .25.
- The **number** of the 1st denomination is 6.
- The **worth** of the 1st denomination is $1.50.

- The **value** of the 2nd denomination (dimes) is .10.
- The **number** of the 2nd denomination is 3.
- The **worth** of the 2nd denomination is 30¢.

The **Total Number** of these two denominations added together is 9 (6 quarters and 3 dimes).

The **Total Worth** of these two denominations added together is $1.80 ($1.50 in quarters and 30¢ in dimes).

8.4 Name The Expression For The Number Of Each Denomination

There are two methods used to name the expression for the number (how many) of each denomination.

Method #1 is when you are given Direct Translation words in the problem. In this event, name the **Totally Unknown** with the variable x, and then build upon that variable with **Direct Translation** to name the other expression. (This method is the same one you have been using in previous sections.)

Method #2 is used when you are *not* given Direct Translation Words. In that event, the problem will always give you the **Total Number** of both denominations added together. Name the expression for the number of one denomination as the variable x, and name the expression for the number of the other denomination as the **Total Number minus x**.

HELPFUL HINT

- Method #2 can be used in any type of word problem when you need to name two different expressions and the only information you have is their total.

	First Expression	Second Expression
Method To Use	x	Total Number $-\ x$
Sample If Total Number Is 50	x	$50 - x$

8.5 Solving The Problem

Step 1
Read Through The Problem

Look for and make note of the three items you need to set up your chart and your equation:

- The **two different denominations** being used so you can determine their **value**.
- **Direct Translation Words** OR a **Total Number** (how many) of the denominations in order to name the expressions for the number of each denomination.
- The **Total Worth** of all the money together.

Money Word Problem Workbook

Step 2
Set Up And Fill In A Chart

Set up a pre-equation chart like the one below to determine the **Denomination Worth.**

- Write the two denominations (such as twenties and fives) in the **Denomination** Column.
- In the **Value** Column, enter the value for each denomination as shown in the chart in 8.2.
- In the **Number** Column, enter the expression for the number of each denomination as explained in 8.4.
- To fill in each **Denomination Worth,** multiply the **Value times the Number**.

Denomination	Value •	Number =	Denomination Worth

Step 3
Set Up An Equation

The equation is the two Denomination Worths from your chart added together and set **equal to the Total Worth** of the money, which is given to you in the problem. The formula is:

$$1^{st} \text{ Denomination Worth } + 2^{nd} \text{ Denomination Worth } = \text{ Total Worth}$$

Step 4
Solve the Equation

Using the method taught by your instructor, solve the equation for the variable x.

Step 5
Make Sure to Answer the Question Being Asked

When you solve the equation, you will find the solution for x, but that might not be the answer to the question. You need to reread the problem and make sure exactly what question is being asked.

It is possible that the amount for the variable x may be your answer. But it may *not* be.
For example, the amount for x may give you the number of five-dollar bills, while the problem may be asking for the number of twenty-dollar bills.

You will have to substitute the solution for x into the original expression for twenty-dollar bills that you will set up in the chart in Step 2 in order to get the correct answer. Always be sure of exactly what the question is.

Word Problem Workbook — Money

EXAMPLES

EXAMPLE 1 Sarah has 6 more fives than twenties in her purse. If she has a total of $105 in her purse, how many fives does Sarah have?

SOLUTION

Step 1 *Read The Problem*

- The two denominations are twenty-dollar bills and five-dollar bills.
- There are Direct Translation Words [more than] given in the problem.
- The Total Worth of the money is $105.

Step 2 *Set Up And Fill In A Chart*

- Write "Twenties" in 1st row and "Fives" in 2nd row of the Denomination Column.
- Fill in the 1st row of the Value Column with "20" as per chart in 8.2.
- Fill in the 2nd row of the Value Column with "5" as per chart in 8.2.
- Use Direct Translation to write expressions for the *number* of each denomination.
- Fill in the 1st row of the Number Column with "x" to represent the number of Twenties.
- Fill in the 2nd row of the Number Column with "$x + 6$" to represent the number of Fives.
- Multiply the Value times the Number to get each Denomination Worth.

Denomination	Value	•	Number	=	Denomination Worth
Twenties	20		x		$20(x)$
Fives	5		$x + 6$		$5(x + 6)$

Step 3 *Set Up The Equation*

- Use the two Denomination Worths from your chart. They are $20(x)$ and $5(x + 6)$.
- The Total Worth is always given in a Money problem. In this example, it is $105.
- Set up equation as 1st Denomination Worth + 2nd Denomination Worth = Total Worth.

$$20(x) + 5(x + 6) = 105$$

Step 4 *Solve The Equation*

- The solution to the equation is $x = 3$

Money   Word Problem Workbook

Step 5 *Answer The Question Asked*

- You have the solution to the equation, but it is NOT the answer to the question.
- The amount of x is the number of Twenties; the problem asks the number of *Fives*.
- You need to use the expression for the number of Fives you named in the chart in Step 2.
- Get the answer by substituting the solution for x (which is 3) into that expression.

> Number of Fives = $x + 6$
> Number of Fives = $3 + 6$
> Number of Fives = 9

Answer: Sarah has 9 five-dollar bills in her purse.

EXAMPLE 2   Consuelo has $1.70 in change consisting of three more dimes than quarters. Find the number of dimes and quarters that she has.

SOLUTION

Step 1 *Read The Problem*

- The two denominations are Dimes and Quarters.
- There are Direct Translation Words [more than] given.
- The Total Worth of the money is $1.70.

Step 2 *Set Up And Fill In A Chart*

- Write "Quarters" in the 1st row and "Dimes" in the 2nd row of the Denomination Column.
- Fill in the 1st row of the Value Column with ".25" as per chart in 8.2.
- Fill in the 2nd row of the Value Column with ".10" as per chart in 8.2.
- Use Direct Translation to write expressions for the *number* of each denomination.
- Fill in the 1st row of the Number Column with "x" to represent the number of quarters.
- Fill in the 2nd row of the Number Column with "$x + 3$" to represent the dimes.
- Multiply the Value times the Number to get each Denomination Worth.

Denomination	Value	•	Number	=	Denomination Worth
Quarters	.25		x		$.25(x)$
Twenties	.10		$x + 3$		$.10(x + 3)$

56

Word Problem Workbook   Money

Step 3 *Set Up The Equation*

- Use the two Denomination Worths from your chart. They are $.25(x)$ and $.10(x + 3)$.
- The Total Worth is always given in a Money problem. In this example, it is $1.70.
- Set up equation as 1st Denomination Worth + 2nd Denomination Worth = Total Worth.

$$.25(x) + .10(x + 3) = 1.70$$

Step 4 *Solve The Equation*

- The solution to the equation is $\boxed{x = 4}$

Step 5 *Answer The Question Asked*

- You have the solution to the equation, but it does not answer the COMPLETE question.
- The amount of x is the number of Quarter; you are also asked for he number of Dimes.
- Use the expression for the number of Dimes you named in the chart in Step 2.
- Get the answer by substituting the solution for x (which is 5) into that expression.

Number of Dimes = $x + 3$
Number of Dimes = $4 + 3$ *Answer*: Consuelo has 4 quarters and 7 dimes.
Number of Dimes = 7

EXAMPLE 3 Jacob has a total of 15 bills worth $120 in his wallet. If he only has five-dollar bills and ten-dollar bills in his wallet, how many fives does he have?

SOLUTION

Step 1 *Read The Problem*

- The two denominations are five-dollar bills and ten-dollar bills.
- There are NO Direct Translation Words given in the problem.
- You ARE given the Total Number of the denominations added together (15).
- The Total Worth of the money is $120.

Money   Word Problem Workbook

Step 2 *Set Up And Fill In A Chart*

- Write "Fives" in the 1st row and "Tens" in the 2nd row of the Denomination Column.
- Fill in the 1st row of the Value Column with "5" as per chart in 8.2.
- Fill in the 2nd row of the Value Column with "10" as per chart in 8.2.
- The Total Number method from Section 8.4 is used to name the Number expressions.
- Fill in the 1st row of the Number Column with "x" to represent the number of fives.
- Fill in the 2nd row of the Number Column with "$15 - x$" to represent the number of tens.
- Multiply the Value times the Number to get each Denomination Worth.

Denomination	*Value* •	*Number* =	*Denomination Worth*
Fives	5	x	$5(x)$
Tens	10	$15 - x$	$10(15 - x)$

Step 3 *Set Up The Equation*

- Use the two Denomination Worths from your chart. They are $5(x)$ and $10(15 - x)$.
- The Total Worth is always given in a Money problem. In this example, it is $120.
- Set up equation as 1st Denomination Worth + 2nd Denomination Worth = Total Worth.

$$5(x) + 10(15 - x) = 120$$

Step 4 *Solve The Equation*

- The solution to the equation is $\boxed{x = 6}$

Step 5 *Answer The Question Asked*

- You have the solution to the equation, and it *is* the answer to the question.
- The value of x is the number of fives. That is what the question is asking.

Answer: Jacob has 6 five-dollar bills in his wallet.

Section 8: Money Exercise Set

1. When Rafael emptied his pockets, he found he had a total of $3.50 in quarters and nickels. If he had 8 more quarters than nickels, how many quarters did Rafael have?

2. A collection of coins consists of dimes and nickels. The number of dimes is two more than the twice the number of nickels. The value of the collection is $2.70. How many dimes are in the collection?

3. Giovanni has 18 more one-dollar bills than five-dollar bills that he collected selling raffle tickets. He sold $78 worth of tickets. How many five-dollar bills did he collect?

4. When the Broward Center of the Performing Arts counted the twenties and fifties in their cash register, they totaled $1720. If there were 16 more twenties than fifties, how many twenties were there?

5. A jar of coins consisting of nickels and dimes totals $11.10. If there are 22 more nickels than twice the amount of dimes, how many dimes and nickels are there?

6. A collection of 57 coins is made up of quarters and dimes. If the collection totals $9.00, how many quarters are there?

7. While selling tickets to the school play, Lee Lai collected $265 in five-dollar bills and ten-dollar bills. If the numbers of tens was 16 less than twice the number of fives, how many five-dollar bills and ten-dollar bills did Lee Lai collect?

8. Mario's Pizzeria made a total of $220 during lunch, consisting of one-dollar bills and five-dollar bills. If the number of fives was 4 less than three times the number of ones, how many one-dollar bills did Mario's Pizzeria have after lunch?

9. Jack had 46 coins in a cigar box consisting of dimes and quarters. If he had a total of $7.90 in the cigar box, how many of each denomination did Jack have?

10. Cheyenne kept quarters and dimes in the change holder of her car to use for tolls on the Sawgrass Expressway. She had 30 coins in the change holder totaling $6.30. How many of each type of coin did Cheyenne have?

11. Irene had a total of 40 bills hidden under her mattress. There were fifty-dollar bills and twenty-dollars bills totaling $1160. How many of each type of bill did Irene have?

12. At the end of each week, José would take the ones and fives out of his wallet to save up for spending money on his vacation. After a year, he had a total of 220 bills that were worth a total of $600. How many of each type of bill did José have?

13. Regina withdrew $270 in fifties and twenties from an ATM. She got 3 less fifties than twenties. How many fifties did Regina get?

Money   Word Problem Workbook

14. Ariel deposited $300 consisting of tens and twenties into her bank account. She deposited 6 more tens than twenties. How many of each bill did Ariel deposit?

15. Suppose you have $15.20 in pennies and dimes. If you have 20 more pennies than 5 times the number of dimes, how many of each coin do you have?

16. Suppose you have $6.05 in nickels and quarters. If you have twice as many quarters as nickels, how many nickels do you have?

17. Jasmine has $1090 in twenties and fifties. If she has 7 less fifties than twice the number of twenties, how many twenties does she have?

18. Aaron purchased $270 worth of computer accessories using ten-dollar bills and five-dollar bills. If he used 6 less fives than tens, how many fives did he use?

19. Rusty saved $980 for a rainy day. He had a total of 60 bills consisting of tens and twenties. How many ten-dollar bills did Rusty save?

20. Susie found $1.95 in nickels and dimes. If she found a total of 25 coins, how many nickels and dimes did she find?

Section 9

Simple Interest

9.1 Explaining Simple Interest Problems

It's easy to recognize a Simple Interest Problem.

- **Two amounts of money** are invested at different rates of interest for one year.
- The different interest **rates** are given as **percents**.
- The amounts of the two investments are *not* given.
- The **total amount of interest** earned from the two investments *is* given and may be stated in various ways.

Sometimes the problem may state, "total annual interest income", other times it may state "total yearly interest from both accounts", "annual simple interest", "annual income of investments", and "total earned income".

There may be still further variations of these phrases. It is not necessary to remember all of them. Just recognize that they all refer to the same thing: **How much money is earned after one year from two different amounts invested at different rates.**

9.2 The Difference Between The Given Money Amounts

There are always **two money amounts** given in a Simple Interest Problem. One of these money amounts will be included in the information about the two amounts of money invested.

This given money amount will either be the **Total Amount Invested** (the two investments added together), or it will be **part of the Direct Translation** that describes one of the amounts invested.

In either case, this given money amount will be used to name the expressions for the two amounts of money invested. These expressions will be entered into a pre-equation chart.

The **other money amount** given in the problem will be the total amount of interest earned from both investments. As explained in 9.1, this amount can be stated in many ways. However, in this section, it will be referred to as the **Total Interest Earned**.

The money amount given for the **Total Interest Earned** is the exact amount that you will use in the setup of the equation.

Simple Interest   Word Problem Workbook

9.3 How To Name The Expressions For The Amounts Invested

As in Section 8, there are two methods used to name the expression for each amount invested. One method is to use Direct Translation.

If Direct Translation Words are not given, the problem will always give the Total Amount Invested. This allows you to use the alternate method: One of the amounts invested as the variable x, and other amount invested as the Total Amount Invested minus x.

	1st Expression	2nd Expression
Alternate Method To Use	x	Total Amount − x
If Total Amount Invested Is 720	x	$720 - x$

9.4 Solving The Problem

Step 1
Read Through The Entire Problem

Look for and make note of the three items you need to set up your chart and your equation:
- The **two different rates** of interest which are in the form of percents.
- **Direct Translation** with a money amount **OR** a **Total Amount Invested** in order to name the expressions for the two amounts invested.
- The **Total Interest Earned**.

Step 2
Fill In The Chart Using The Interest Formula

Set up a pre-equation chart like the one below. This will give you the two **Interest Terms** that you need to use in your equation.

The formula used is "**Interest (I) = principal (p) • rate (r)**".
- The "**rate**" is the percentage that is given in the problem.
- The "**principal**" refers to each amount of money invested.
- The "**interest**" is the amount earned in one year from the investment. This is determined by **multiplying the rate times the principal** (amount invested) for each investment.

r	•	p	=	I

Each row in the chart represents one investment. Enter in the chart the information given in the problem that refers to the **rate** (percent) and **principal** (amount invested) of each investment.

The **rate** is given in the problem, but you will need to name expressions for the two amounts of money invested. *Use one of the methods as explained in 9.3.*

The Interest earned for each investment is not given in the word problem. You will get the expression to fill in the Interest portion of the chart by using the **Interest formula** and multiplying together the rate and principal you enter.

These two Interest expressions will be needed to set up the equation.

NOTE: The full formula for interest is "$I = p \cdot r \cdot t$". The "t" represents "time". Since the time in Simple Interest Problems is always 1 (for one year), "t" is not used.

HELPFUL HINT

- Remember that you cannot multiply by a percent. Be sure to **convert the percent to a decimal** before entering it into the rate column of the chart.

- When using Direct Translation to get the expressions for the amounts invested, be careful to match up each amount invested with its corresponding rate. The word "**at**" is the word to look for in the problem that **connects an amount to its matching rate**.

Step 3

Set Up An Equation

Not everything from your chart goes into the equation. What you will need is the two **interest expressions.**

The **Interest expressions** which represent the interest earned for each investment have already been determined and entered in your chart. Set up the two Interest expressions equal to the "**Total Interest Earned**", the amount that is given to you in the problem.

$$1^{st} \text{ Interest} + 2^{nd} \text{ Interest} = \text{Total Interest Earned}$$

Step 4
Solve the Equation

Using the method taught by your instructor, solve the equation for the variable.

Step 5
Make Sure to Answer the Question Being Asked

In Simple Interest Problems, as in other word problems, you need to make sure exactly what question is being asked. It is possible that the value for the variable x may be your answer. But it may *not* be.

For example, the value for x may be the amount of money invested at 6%, and the question wants to know the amount of money invested at 8%. To get the correct answer, look at your pre-equation chart and find the expression for the amount invested that corresponds to the 8% rate. Substitute the solution for x into that expression to get the correct answer.

EXAMPLES

EXAMPLE 1 Zach Thomas has some money invested at 5%, and $5000 more than that amount invested at 9%. His total annual interest income is $1430. Find the amount invested at 9%.

SOLUTION

Step 1 *Read The Problem*

- The two rates of interest are 5% and 9%.
- Direct Translation and a money amount are given in the problem.
- These will be used to name the expressions for the two amounts of money invested.
- The Total Interest Earned is $1430.

Word Problem Workbook
Simple Interest

Step 2 *Set Up And Fill In A Chart*

- Write in the investment percentage in the Investment Column.
- Fill in the 1st row of the Rate Column with ".05", the decimal equivalent of 5%.
- Fill in the 2nd row of the Rate Column with ".09", the decimal equivalent of 9%.
- Use Direct Translation to write expressions for the amount of each investment.
- Fill in 1st row of the Principal Column with "x" to represent the amount at 5%.
- Fill in 2nd row of the Principal Column with "$x + 5000$" to represent the amount at 9%.
- For each investment, multiply the Rate times the Principal to get its Interest Expression.

Investment	Rate •	Principal (Amount)	=	Interest
Amount at 5%	.05	x		.05(x)
Amount at 9%	.09	$x + 5000$		.09($x + 5000$)

Step 3 *Set Up The Equation*

- Use the two Interest expressions from your chart. They are .05(x) and .09($x + 5000$).
- The Total Interest Earned is always given in the problem. In this example, it is $1430.
- Set up equation as Interest Expression + Interest Expression = Total Interest Earned.

$$.05(x) + .09(x + 5000) = 1430$$

Step 4 *Solve The Equation*

- The solution to the equation is $\boxed{x = 7000}$

Step 5 *Answer The Question Asked*

- You have the solution to the equation, but it is NOT the answer to the question.
- x is the amount invested at 5%; the problem asks for the amount invested at 9%.
- You need to use the expression for the 9% investment you named in the chart in Step 2.
- Get the answer by substituting the solution for x (which is 7000) into that expression.

Amount at 9% = $x + 5000$
Amount at 9% = 7000 + 5000
Amount at 9% = 12,000

Answer: The amount invested at 9% is $12,000.

Simple Interest   *Word Problem Workbook*

EXAMPLE 2   Whitney invested part of her $25,000 advance in a savings account at 7% annual simple interest and the rest in a mutual fund at 8% annual simple interest. If her total yearly interest from both accounts was $1900, find the amount invested at each rate.

SOLUTION

Step 1   *Read The Problem*

- The two rates of interest are 7% and 8%.
- The Total Amount Invested is given in the problem. It is $25,000.
- This will be used to name the expressions of the individual amounts of money invested.
- The Total Interest Earned is $1900.

Step 2   *Set Up And Fill In A Chart*

- Write in the investment percentage in the Investment Column.
- Fill in the 1st row of the Rate Column with ".07", the decimal equivalent of 7%.
- Fill in the 2nd row of the Rate Column with ".08", the decimal equivalent of 8%.
- Use the Total Amount method to write expressions for the amount of each investment.
- Fill in 1st row of the Principal Column with "x" to represent the amount at 7%.
- Fill in 2nd row of the Principal Column with "$25,000 - x$" to represent the amount at 8%.
- For each investment, multiply the Rate times the Principal to get its Interest Expression.

Investment	Rate •	Principal (Amount)	=	Interest
Amount at 7%	.07	x		$.07(x)$
Amount at 8%	.08	$25,000 - x$		$.08(25,000 - x)$

Step 3   *Set Up The Equation*

- Use the two Interest expressions from your chart. They are $.07(x)$ and $.08(25,000 - x)$.
- The Total Interest Earned is always given in the problem. In this example, it is $1900.
- Set up your equation as Interest Expression + Interest Expression = Total Interest Earned.

$$.07(x) + .08(25,000 - x) = 1900$$

Word Problem Workbook

Simple Interest

Step 4 *Solve The Equation*

- The solution to the equation is $\boxed{x = 10{,}000}$

Step 5 *Answer The Question Asked*

- You have the solution to the equation, but it is not the complete answer to the question.
- x is the amount invested at 7%; the problem also asks for the amount invested at 8%.
- You need to use the expression for the 8% investment you named in the chart in Step 2.
- Get the answer by substituting the solution for x (which is 10,000) into that expression.

$$\boxed{\begin{array}{l}\text{Amount at 8\%} = 25{,}000 - x \\ \text{Amount at 8\%} = 25{,}000 - 10{,}000 \\ \text{Amount at 8\%} = 15{,}000\end{array}}$$

Answer: The amount invested at 7% is $10,000.
The amount invested at 8% is $15,000.

Section 9: Simple Interest Exercise Set

1. Clay Aiken has an annual interest income of $3390 from two investments. He has $10,000 more invested at 8% than he has invested at 6%. Find the amount invested at 6%.

2. Frank Gore has some money invested at 5%, and $5000 more than that amount invested at 9%. His total annual interest income is $1430. Find the amount invested at 5%.

3. Cristian Vega received an inheritance. He invested some of the inheritance at 9% and $3500 more than that amount at 10%. If he earns $1490 in annual interest from the two investments, find the amount he invested at 9%.

4. Marilyn Milian earns a $17,000 bonus from her company. She invests part of the money at 9% and the balance at 11%. If the annual interest for the two investments is $1670, find the amount invested at each rate.

5. Larry Coker has $1000 more invested at 9% than he has invested at 11%. If the annual income for the investments is $1290, how much he has invested at each rate?

6. Ken Dorsey received a $100,000 signing bonus. He invested some of the money in a Mutual Fund at 8% and the rest of the money in Savings Bonds at 5%. If his total yearly interest earned was $6800, how much did Ken Dorsey invest in the Mutual Fund at 8%?

7. Hillary B. Smith invested her holiday bonus check of $5,000 into two different savings accounts. She invested some at 3% and the rest at 7%. If her total yearly interest was $290, how much did she invest at 3%?

8. The CEO of Hammer Industries invested $75,000 in stocks and bonds. The stocks had an interest rate of 11% and the bonds had an interest rate of 6%. If her total earned income for one year was $6750, how much was invested at 11%?

9. Blair invested some money at 9% annual simple interest and $250 more than that amount at 10% annual simple interest. If her total yearly interest was $101, how much was invested at each rate?

10. Hilda invested $15,000 from which she earns an annual income of $1620. She invested part of the money in a money market at 9% and the rest of the money in a mutual fund at a rate of 12%. How much did Hilda invest in the money market?

11. Sergei has an account with his bank that pays 3% interest, and an account with his credit union that pays 5%. He has $1000 more invested at 5% than twice the amount he has invested at 3%. If his interest earned was $245 last year, how much did Sergei have invested in his bank account at 3%?

Word Problem Workbook   Simple Interest

12. Vernon Duke invests some money in a savings account at 6% simple interest and 600 more than that amount in stocks at 11% simple interest. If his yearly interest income was $185, how much did he have invested at each rate?

13. Missy Elliott invests some money in bonds at 8.5% simple interest and $2000 more than twice that amount in a retirement account at 10% simple interest. If her total annual income is $1625, how much did she invest in bonds 8.5%?

14. Erika Slezak invests some money in stocks at 7.5% simple interest and $1000 less than twice that amount in savings bonds at 9.5%. If her yearly income earned is $2025, how much did she invest in savings bonds at 9.5%?

15. Brooke and Jeff received a total of $12,000 in wedding gifts. They invested some in her savings account at 6.5% simple interest and the rest in his savings account at 4.5% simple interest. If they received an annual income of $680, how much did they invest in each account?

16. Phil McGraw invested a total of $70,000 in two different money markets. One yielded 11.5% simple interest and the other yielded 15.5% simple interest. If he earned an annual income of $10,050, how much was invested in the money market that yielded 11.5%?

17. Fatima invested some money at 6% and $500 more than that amount at 8%. If the total income earned was $320, how much did she invest at each rate?

18. Karenna has an annual interest income of $1380. She has $6000 more invested at 11% than she does at 7%. Find the amounts Karenna invested at each rate.

19. Matchbox Twenty got an advance of $80,000 after signing with Sony Records. They invested some of this money at 8% and the rest at 10%. If their yearly earned income was $7300, how much did Matchbox Twenty invest at 10%.

20. Emanuel invested a total amount of $8000 into two separate mutual funds. One fund yields 5% while the other yields 7%. If the income earned in one year on the two investments is $510, how much did Emanuel invest at 5%?

Section 10

Mixture

In Mixture Problems, there are two substances that are the *same item* but with *different strengths*. Varying amounts of these two substances are mixed together to form a **final mixture** of the same substance with still another strength.

For example, if you mixed 1cup of chocolate milk that was 23% Hershey's syrup with 2 cups of chocolate milk that was 8% Hershey's syrup, you would have a final mixture in the amount of 3 cups of chocolate milk that was 13% Hershey's syrup.

10.1 Explaining Mixture Problems

In mixture problems, the strength of the substances are given as **percentages**. All three percentages of the substances are always given -- the percents of the two substances mixed together and the percent of the final mixture.

What you will be solving for in the problem is one or more amounts of the substances you mix together in order to obtain the desired percentage of the final mixture.

One of the three substance amounts is always given. It may be the amount of one of the substances that will be mixed together, or it may be the total amount of the final mixture.

HELPFUL HINT

- When the percent for a substance is not given, it will state in the problem that the substances is **pure**. That means that it is 100% of that substance and not diluted in any way. In that event, you use **100%** as the percentage for that substance.

10.2 Determining Correct Substance

- It is EXTREMELY important to match the percent of each substance with its corresponding amount. The key word to look for is "**of**". The **amount** stated *before* the word "**of**" connects to the **percent** given *after* the word "of".

- For example, in the statement "A pharmacist has **2 liters** *of* a solution containing **30%** alcohol", the word "of" connects the amount "2 liters" with "30%". Therefore, you know that information belongs to the same substance. In this particular case, you have been given both the percent and the amount of the substance.

- Another example is the statement "**How many** cubic centimeters *of* a **25%** antibiotic solution", the phrase "how many" (which has to do with amount) connects the word "of" with 25%. Therefore, you know this information belongs to the same substance. In this example, you have the percent, but the amount is unknown and will need an expression.

- To determine which of the substances is the Final Mixture, you need to look for information that follows such phrases as "**to get, to obtain, to have, to make**" and others phrases that have the same type of meaning.

- The information that is stated after such a phrase is the information that you will be given about the **final mixture**. You will always be given a percent for the final mixture. The *amount* of the final mixture may or may not be given.

10.3 How To Express The Amounts

Before you can begin to solve the problem, you need to **determine the amount of each substance** that is mixed together. In the problem, you may be given the amount of one of these substances. In that event, use the variable x for the amount of the other substance.

In order to get the expression to represent the total amount of the final mixture, **add together the amounts of the two substances**. For example, if the amount of the 1st substance is given as 3 quarts, the variable x would be used for the 2nd unknown amount of substance, and the total amount of the final mixture would be expressed as "$3 + x$", as in the table below.

1st Substance Amount	+	2nd Substance Amount	=	Final Mixture Total Amount
3		x		$3 + x$

Sometimes, instead of being given the amount of one of the substances that is mixed together, the only amount you are given is the **Total Amount of the final mixture**. In that event, use the alternate **Total Amount method** to find expressions for the two amounts of substances that are mixed together: Use the variable x for one of the substances and then use the Total Amount of the Final Mixture minus x for the other substance.

For example, if you are given 14 gallons as the amount of the Final Mixture, name the amount of the 1st substance with the variable x, and name the amount of the 2nd substance "$14 - x$".

1st Substance Amount	+	2nd Substance Amount	=	Final Mixture Total Amount
x		$14 - x$		14

Mixture

10.4 Solving The Problem

NOTE: The substance type and unit of measurement are not used in solving the problem.

Step 1
Read The Problem And Match The Substances

- Take note of the **percentages** given and to which amount they belong.
- Find the information that belongs to the **final mixture substance**.
- See if the given amount is for a substance you mix or for the total amount of the final mixture so you know how to name the amounts as explained in 10.3.

Step 2
Set Up And Fill In A Chart

Set up a pre-equation chart like the one below. This is how you determine the three Substance Terms that you need in the next step in order to set up your equation.

Substance	Percent •	Amount	= Substance Term
1st Substance			
2nd Substance			
Final Mixture			

The rows in the chart represent the two substances that are mixed together and the substance that will be the final mixture. In the **Percent Column**, enter the percentage for each substance as given in the problem.

In the **Amount Column**, enter the names of the expression for the two substances and for the final mixture **as explained in 10.3**.

To get the expression to fill in the column for the **Substance Terms** portion of the chart, **multiply the Percent times the Amount.**

HELPFUL HINT

- If your instructor has no objection, use this shortcut: Instead of changing the percents to decimal, drop the percent symbol and use the whole number. **This can only be done in Mixture problems** (because every term in the equation has a percent).

- It is extremely important to Double Check to verify that you have the percent paired up with its corresponding amount as explained in 10.2.

Word Problem Workbook   Mixture

Step 3
Set Up An Equation

Once you have correctly set up and filled in your chart, it is very easy to set up your equation. Set up the three Substance Terms from the chart as follows:

Set up the Substance Term that represents the 1st substance and add to it the Substance Term that represents the 2nd substance. Set these two terms equal to the Substance Term that represents the Final Mixture.

> 1st Substance Term + 2nd Substance Term = Final Mixture Substance Term

Step 4
Solve the Equation

Using the method taught by your instructor, solve the equation for the variable.

Step 5
Make Sure to Answer the Question Being Asked

In Simple Interest Problems, as in other word problems, you need to make sure exactly what question is being asked. It is possible that the value for the variable x may be your answer. But it may *not* be.

For example, the value for x may be the amount of one of the substances you mix together, and the question wants to know the amount of the final mixture substance.

To get the correct answer, look at your pre-equation chart and find the expression for the amount that will answer the question. Substitute the solution for x into that expression to get the correct answer.

Mixture Word Problem Workbook

EXAMPLES

EXAMPLE 1 How many quarts of a 25% antibiotic solution should be added to 10 quarts of a 60% antibiotic solution in order to get a 30% antibiotic solution?

SOLUTION

Step 1 *Read The Problem*

- 25% goes with an unknown amount of a substance that is mixed.
- 60% goes with 10 quarts of a substance that is mixed.
- The final mixture substance is 30%.
- The amount given, 10 quarts, is for a substance that is mixed.

Step 2 *Set Up And Fill In A Chart*

- Fill in 25, 60, and 30 in the percent column next to their corresponding substances.
- As per 10.3, name the expressions for all the amounts and fill in the amounts column.
- The 1st substance (25%) was unknown, so the amount is x.
- The amount of the 2nd substance (60%) was given in the problem. It is 10.
- The amount of the Final Mixture (30%) is the total of the other two substances, $x + 10$.
- Multiply the Percent times the Amount to get each Substance Term.

Substance	Percent •	Amount =	Substance Term
1st Substance	25	x	$25(x)$
2nd Substance	60	10	$60(10)$
Final Mixture	30	$x + 10$	$30(x + 10)$

Step 3 *Set Up The Equation*

- Use the three Substance Terms from your chart. They are $25(x)$, $60(10)$, and $30(x + 10)$.
- Set up your equation as 1st Substance Term + 2nd Substance Term = Final Mixture Term.

$$25(x) + 60(10) = 30(x + 10)$$

Word Problem Workbook   *Mixture*

Step 4 *Solve The Equation*

- The solution to the equation is $\boxed{x = 60}$

Step 5 *Answer The Question Asked*

- The question asks for the amount of the 25% solution substance.
- x is the amount of the 25% solution. You are done. You have the correct answer.

    Answer: Add 60 quarts of the 25% solution.

EXAMPLE 2   How much pure alcohol must be added to 2 liters of a solution containing 30% alcohol to obtain a solution containing 44% alcohol?

SOLUTION

Step 1 *Read The Problem*

- "Pure" means 100%, so 100% goes with an unknown amount of a mixing substance.
- 30% goes with 2 liters of a mixing substance.
- The final mixture substance is 44%.
- The amount given, 2 liters, is for a substance that is mixed.

Step 2 *Set Up And Fill In A Chart*

- Fill in 100, 30, and 44 in the percent column next to their corresponding substances.
- As per 10.3, name the expressions for all the amounts and fill in the amounts column.
- The 1st substance (100%) was unknown, so the amount is x.
- The amount of the 2nd substance (30%) was given in the problem. It is 2.
- The amount of the Final Mixture (44%) is the total of the two mixing substances, $x + 2$.
- Multiply the Percent times the Amount to get each Substance Term.

Substance	Percent •	Amount	= Substance Term
1st Substance	100	x	$100(x)$
2nd Substance	30	2	$30(2)$
Final Mixture	44	$x + 2$	$44(x + 2)$

Mixture *Word Problem Workbook*

Step 3 *Set Up The Equation*

- Use the three Substance Terms from your chart. They are $100(x)$, $30(2)$, and $44(x + 2)$.
- Set up your equation as 1st Substance Term + 2nd Substance Term = Final Mixture Term.

$$100(x) + 30(2) = 44(x + 2)$$

Step 4 *Solve The Equation*

- The solution to the equation is $\boxed{x = 0.5}$

Step 5 *Answer The Question Asked*

- The question asks for the amount of the pure (100%) solution substance.
- x is the amount of the 100% solution. You are done. You have the correct answer.

   Answer: Add .5 liter (one-half of a liter) of the pure solution.

EXAMPLE 3 A chef wants to mix a 60% sugar solution with a 30% sugar solution to obtain 10 pints of a 51% sugar solution. How much of the 30% solution will the chef use?

SOLUTION

Step 1 *Read The Problem*

- 60% goes with an unknown amount of a substance that is mixed.
- 30% goes with an unknown amount of a substance that is mixed.
- The final mixture substance is 51%.
- The amount given, 10 pints, is for the total amount of the final mixture.

Step 2 *Set Up And Fill In A Chart*

- Fill in 60, 30, and 51 in the percent column next to their corresponding substances.
- As per 10.3, name the expressions for all the amounts and fill in the amounts column.
- The amount of the Final Mixture (51%) is the only amount given in the problem. It is 10.
- The 1st substance (60%) is unknown, so the amount is x.
- The 2nd substance (30%) is determined by the alternate Total Amount Method. It is $10 - x$.
- Multiply the Percent times the Amount to get each Substance Term.

Word Problem Workbook — Mixture

Substance	Percent •	Amount	=	Substance Term
1ˢᵗ Substance	60	x		$60(x)$
2ⁿᵈ Substance	30	$10 - x$		$30(10 - x)$
Final Mixture	51	10		$51(10)$

Step 3 *Set Up The Equation*

- Use the three Substance Terms from your chart. They are $60(x)$, $30(10 - x)$, and $51(10)$.
- Set up your equation as 1ˢᵗ Substance Term + 2ⁿᵈ Substance Term = Final Mixture Term.

$$60(x) + 30(10 - x) = 51(10)$$

Step 4 *Solve The Equation*

- The solution to the equation is $x = 7$

Step 5 *Answer The Question Asked*

- You have the solution to the equation, but it is NOT the answer to the question.
- The value of x is the 60% substance; the problem asks for the 30% substance.
- You need to use the expression for the 30% substance that you named in Step 2.
- Get the answer by substituting the solution for x (which is 7) into the expression.

30% Substance = $10 - x$
30% Substance = $10 - 7$
30% Substance = 3

Answer: The chef needs to use 3 pints of the 30% solution.

Section 10: Mixture Exercise Set

1. How many liters of a 25% salt solution must be added to 20 liters of a 12% solution to get a solution that is 20% salt?

2. How much of an alloy that is 20% copper should be mixed with 200 ounces of an alloy that is 50% copper in order to get an alloy that is 30% copper?

3. How many pounds of a metal containing 35% nickel would be needed to melt and mix with another metal containing 65% nickel to get 50 pounds of a metal containing 50% nickel?

4. How many gallons of an 18% pesticide solution must be added to 92 gallons of a 51% pesticide solution to obtain a 41% pesticide solution?

5. How many liters of pure salt must be added to 15 liters of a 40% salt solution to obtain a 60% salt solution?

6. How many gallons of pure anti-freeze must be mixed with 30 gallons of 15% anti-freeze to get a mixture that is 40% anti-freeze?

7. A pharmacist mixes a 24% iodine solution with a 64% iodine solution to get 160 milliliters of a 43% iodine solution. How much of the 24% iodine solution did she use?

8. A hairdresser mixes a 30% peroxide solution with a 10% peroxide solution to get 4 cups of 20% peroxide solution. How much of the 10% peroxide solution did she use?

9. A jeweler melted a 50% gold metal with a 20% gold metal to get 6 ounces of a 40% gold metal. How much of the 50% gold metal did he use?

10. A lab technician needs to mix a 45% phosphorus solution with an 18% phosphorus solution to obtain 12 milliliters of a 36% phosphorus solution. How many milliliters of the 18% phosphorus solution will the lab technician need?

11. How many pints of a 10% lemon juice marinade must be mixed with 20 pints of a 60% lemon juice marinade to get a 30% lemon juice marinade?

12. How much of a 60% salt solution must be mixed with 108 liters of an 80% salt solution to get a 75 % salt solution?

13. How many cubic inches of a candy bar that is 45% dark chocolate must be melted and mixed with 18 cubic inches of a candy bar that is 20% dark chocolate to obtain a candy bar that is 30% dark chocolate?

14. How much of a 20% alcohol solution must be mixed with 20 liters of a 5% alcohol solution in order to get a 10% alcohol solution?

15. How many tablespoons of pure butter would need to be mixed with 60 tablespoons of a 50% butter spread to get a 70% butter spread?

16. How many pints of pure orange juice must be added to 6 pints of a 25% orange juice drink to get a 55% orange juice drink?

17. How many cups of pure sour cream must be mixed with 12 cups of a 40% sour cream dip to get a 60% sour cream dip?

18. How much pure sodium must be mixed with 50 quarts of a 65% sodium solution to get a 75% sodium solution?

19. How many ounces of a 20% talcum powder should be mixed with 50 ounces of a 60% talcum powder to obtain a 30% talcum powder.

20. A chemistry student must mix a 5% iodine solution with a 12% iodine solution. How much of each must he use to obtain 70 liters of an 8% iodine solution.

Section 11

Geometric Area And The Pythagorean Theorem

Step 1
Read Through The Entire Problem

Determine which type of problem it is in order to know which formula to use to solve the problem. Is it a **Geometric Area** problem, or is it a **Pythagorean Theorem** problem?

In a Geometric Area problem, the word **"area"** will be stated. In a Pythagorean Theorem problem, it will state **"hypotenuse"** and/or **"right triangle"**. Look for Direct Translation words.

Step 2
Name The Expressions

In both Geometric Area and Pythagorean Theorem problems, you need to use Direct Translation to name the expressions you need for the formulas. As in previous sections, keep in mind you will need to name the Totally Unknown with a variable, and then use Direct Translation in order to build on that variable and name the other expressions.

In a Geometric Area problem, you will need to name expressions for both the **length** and the **width** of a rectangle.

In a Pythagorean Theorem problem, you will need to name expressions for the **three sides of a right triangle**. These three sides will generally be referred to as the **shorter leg, the longer leg, and the hypotenuse.**

Step 3
Set Up An Equation By Substituting The Expressions Into The Appropriate Formula

In Geometric Area, the value for the area is given; you need to name expressions for the length and width. In Pythagorean Theorem, you need to name expressions for all three sides.

Problem Type	Geometric Shape	Formula
Geometric Area	Rectangle	$LW = A$
Pythagorean Theorem	Right Triangle	$a^2 + b^2 = c^2$

Word Problem Workbook *Geometric Area And The Pythagorean Theorem*

> *HELPFUL HINTS*
>
> - With the **Geometric Area formula**, it may be easier to solve the equation if you set it up as $WL = A$, if the width (W) is the monomial (single term). $L \cdot W$ is the same as $W \cdot L$.
>
> - When setting up the formula for the **Pythagorean Theorem**, it may be helpful to think of the formula as $(\text{leg})^2 + (\text{leg})^2 = (\text{hypotenuse})^2$.

Step 4
Solve the Equation

Using the method taught by your instructor, solve the equation for the variable. Keep in mind when solving a Geometric Area or a Pythagorean Theorem Problem, you normally get two solutions to the equation. If one of the solutions is negative, **eliminate** it because the length of any side of a shape **cannot be negative**.

> **NOTE**: If both of the solutions are positive, you must substitute each of them into the expressions named in Step 2. Whichever solution makes the value of any expression negative must be eliminated.

Step 5
Make Sure to Answer the Question Being Asked

As in previous sections, when you solve the equation you will find the value for x, but that might not be the answer to the question. You need to reread the problem and make sure exactly what question is being asked.

It is possible that the value for the variable x may be your answer. But it may *not* be.

For example, the value for x may give you the *width* of a rectangle, while the problem may be asking for the *length* of the rectangle.

You will have to substitute the value for x into the original expression for the length that you will set up in Step 2 in order to get the correct answer. Always be sure of exactly what the question is.

Geometric Area And The Pythagorean Theorem *Word Problem Workbook*

EXAMPLES

EXAMPLE 1 The length of a rectangle is 8 feet shorter than three times its width. If the area of the rectangle is 35 square feet, what is the length of the rectangle?

SOLUTION

Step 1 *Read The Problem*

- Seeing the word "area" verifies that it is a Geometric Area problem.
- The problem contains Direct Translation Words [is, shorter than, three times].

Step 2 *Name The Expressions*

- You are given information about the length; it is 8 feet shorter than three times the width.
- You are not given information about the width, so the width is the Totally Unknown.
- Use x for the width. Use Direct Translation to name an expression for the length.
- The area is always given. It is 35.

$$\text{Width } (W) = x$$
$$\text{Length } (L) = 3x - 8$$

Step 3 *Set Up The Equation*

- The width is the monomial, the single term.
- As per Step 3 Helpful Hint, use the formula $WL = A$ instead of $LW = A$.
- The area is 35, so substitute the number 35 for A.
- Substitute the expressions you named in Step 2 for L and W into the formula.

$$x(3x - 8) = 35$$

Step 4 *Solve The Equation*

- The solutions to the equation are $x = -\frac{7}{3}, x = 5$
- One of the solutions is negative.
- Eliminate the negative solution because the length of a side of a shape cannot be negative.
- The correct solution is $x = 5$

Word Problem Workbook   Geometric Area And The Pythagorean Theorem

Step 5 *Answer The Question Asked*

- You have your solution to the equation, but it is NOT the answer to the question.
- The value of x is the size of the width; the problem asks for the length.
- You need to use the expression for the length that you named in Step 2.
- Get the answer by substituting the solution for x (which is 5) into the expression.

$$L = 3x - 8$$
$$L = 3(5) - 8$$
$$L = 7$$

Answer: The length of the rectangle is 7 feet.

EXAMPLE 2 The hypotenuse of a right triangle is one inch longer than the longer leg. The shorter leg is 7 inches shorter than the longer leg. Find the length of the hypotenuse.

SOLUTION

Step 1 *Read The Problem*

- Seeing the word "right triangle" verifies that it is a Pythagorean Theorem problem.
- The problem contains Direct Translation Words [is, longer than, shorter than].

Step 2 *Name The Expressions*

- You are given information about the hypotenuse; it is 1 inch longer than the longer leg.
- You are given information about the shorter leg; it is 7 inches shorter than the longer leg.
- You are not given information about the longer leg, so that is the Totally Unknown.
- Use Direct Translation to name expressions for the hypotenuse and the shorter leg.

Step 3 *Set Up The Equation*

- Use the formula $a^2 + b^2 = c^2$ (Or $\text{Leg}^2 + \text{Leg}^2 = \text{Hypotenuse}^2$).
- Substitute the expressions you named for each leg and the hypotenuse into the formula.

$$x^2 + (x-7)^2 = (x+1)^2$$

Geometric Area And The Pythagorean Theorem   Word Problem Workbook

Step 4 *Solve The Equation*
- The solutions to the equation are $x = 12, \ x = 5$
- When both solutions are positive, substitute each one into the expressions in Step 2.
- $x = 5$ is eliminated because it makes the expression for the Shorter Leg negative.
- The only solution you can use is $x = 12$

Step 5 *Answer The Question Asked*

- You have your solution to the equation, but it is NOT the answer to the question.
- The value of x is the size of the longer leg; the problem asks for size of the hypotenuse.
- You need to use the expression for the hypotenuse you named in Step 2.
- Get the answer by substituting the solution for x (which is 12) into the expression.

$$\text{Hypotenuse} = x + 1$$
$$\text{Hypotenuse} = 12 + 1$$
$$\text{Hypotenuse} = 13$$

Answer: The length of the hypotenuse is 13 inches.

Section 11: Geometric Area & Pythagorean Theorem Exercise Set

1. The width of a rectangle is 7 inches shorter than its length. The area of the rectangle is 44 square inches. Find the length and width of the rectangle.

2. The width of a rectangle is 2 yards shorter than its length. The area of the rectangle is 48 square yards. Find the length and width of the rectangle.

3. The length of a rectangle is 5 feet longer than its width. The area of the rectangle is 66 square feet. Find the length of the rectangle.

4. The length of a rectangle is 4 inches longer than its width. The area of the rectangle is 96 square inches. Find the width of the rectangle.

5. The length of a rectangular picture frame is twice its width. If the area of the frame is 50 square inches, what are the dimensions of the picture frame?

6. The length of a rectangular rug is 3 feet shorter than twice its width. The area of the rug is 54 square feet. What are the dimensions of the rug?

7. The area of a rectangular pool cover is 15 square yards. If the length of the pool cover is one yard shorter than twice its width, what are the dimensions of the pool cover?

8. The area of a rectangular tabletop is 8 square feet. If the length of the tabletop is twice its width, what are the dimensions of the tabletop?

9. The length of a rectangle is 2 feet longer than twice its width. If the area of the rectangle is 40 square feet, what is the length of the rectangle?

10. The length of a rectangle is 3 inches longer than 3 times its width. If the area of the rectangle is 36 square inches, what is the length of the rectangle?

11. The longer leg of a right triangle is 7 feet longer than the shorter leg. The hypotenuse is 13 feet. Write the length of all three sides.

12. The longer leg of a right triangle is 7 feet longer than the shorter leg. The hypotenuse is 17 feet. Write the length of all three sides.

13. The shorter leg of a right triangle is 2 inches shorter than the longer leg. The hypotenuse is 10 inches. How long is the shorter leg?

14. The shorter leg of a right triangle is 14 yards shorter than the longer leg. The hypotenuse is 26 yards. How long is the shorter leg?

15. The hypotenuse of a right triangle is 4 meters longer than the longer leg. The shorter leg is 4 meters shorter than the longer leg. Find all three sides.

Geometric Area And The Pythagorean Theorem *Word Problem Workbook*

16. The longer leg of a right triangle is 2 feet shorter than twice the shorter leg. The hypotenuse is 2 feet longer than twice the shorter leg. How long is the hypotenuse?

17. The hypotenuse of a right triangle is 4 yards longer than 3 times the shorter leg. The longer leg is 3 yards longer than three times the shorter leg. Find the length of all three sides.

18. The hypotenuse is 5 inches shorter than 6 times the shorter leg. The longer leg is 5 inches longer than 5 times the shorter leg. Find the length of all three sides.

19. A 10 foot ladder is leaning against the side of a building. The distance from the bottom of the building to the top of the ladder is 2 more feet than the distance from the bottom of the building to the base of the ladder. What is the distance from the bottom of the building to the base of the ladder? (Hint: It would be helpful to draw a sketch.)

20. Two cars left the same intersection. One car traveled to the north. The other car traveled to the east. The car that went east drove 7 miles further than the car that traveled to the north. At that point the two cars were 13 miles away from each other. How far did each car travel? (Hint: It would be helpful to draw a sketch.)

Section 12

Motion – Rational Equations

These types of problems are the same concept as the Motion Problems in Section 7, but they involve extra elements and different components.

12.1 Explaining Motion – Rational Equations

In this section, there may be two objects in motion but most often there is just one, traveling at different rates and different distances. The times may be the same, or the time may be given as a total.

Another component is that the rates of the objects are sometimes affected by outside influences such as the speed of **water current** or the speed of the **wind**.

12.2 How To Express The Rates

Before you can begin to solve the problem, you need to take note of the **rates** of any or each object in motion. You name the expressions for the rates in one of two ways. One way is Direct Translation with which you are already familiar.

When Direct Translation is not given, there will be information given about the water current or the wind. This information will be used to determine expressions for the rates.

When water current is involved, the word "**current**" may be stated in the problem, or words such as "**downstream**" or "**upstream**".

The current affects the rate of an object because when an object is traveling downstream, it means it is going with the current. In that case, the object is being pushed along by the water and would be traveling at a faster speed. However, when an object is traveling upstream, it means it is fighting *against* the water and would be traveling at a slower speed.

The chart below will give you the information you need to name expressions for the rate when current is involved. The variable x represents the **rate of an object in still water** without the influence of any current. The variable c represents the **speed of the current**.

Path Of Object	What It Means	What It Does	If x Is Rate In Still Water	Expression To Use
Downstream	With Current	Increases Rate	Add Current To Rate	$x + c$
Upstream	Against Current	Slows Down Rate	Deduct Current From Rate	$x - c$

Motion – Rational Equations *Word Problem Workbook*

When wind speed is involved, the words "**speed of the wind**" and "**wind blowing**" may be stated in the problem, or you may see words such as "**tailwind**" or "**headwind**".

When an object is traveling with a tailwind, it means that the object has the wind behind it, and it is being pushed along by the pressure of the wind. Therefore, the object would be traveling at a faster speed.

In turn, an object with a headwind would be traveling against the wind, and would be held back by air resistance. Therefore, the object would be traveling at a slower speed.

The chart below will give you the information you need to name expressions for the rate when the speed of the wind is involved. The variable x represents **the rate of an object in still air**, without the influence of any wind. The variable w represents the **speed of the wind**.

Direction Of Wind	What It Means	What It Does	If x Is Rate In Still Air	Expression To Use
Tailwind	With The Wind	Increases Rate	Add Wind To Rate	$x + w$
Headwind	Against The Wind	Slows Down Rate	Deduct Wind From Rate	$x - w$

12.3 Solving The Problem

Step 1
Read Through The Entire Problem

To fill in your pre-equation chart, you need to look for and take note of:

- The distances given.
- Is there Direct Translation given for the rates?
- Is there information about influences on the rates as explained in 12.2?

To use the correct equation, you must know if the **time given** is the *same* for both objects, or if you are given a *total time*.

Word Problem Workbook   Motion – Rational Equations

Step 2
Set Up And Fill In The Chart

Set up a pre-equation chart like the one below. This will be used to determine the expression for Time that you will need for the equation.

The formula used is a variation of the distance formula, which is manipulated in order to change it to a formula that will give a rational expression for the time. This variation is **Distance divided by Rate equals Time.**

$$\text{Time } (t) = \frac{\text{Distance } (d)}{\text{Rate } (r)}$$

d	r	t

Enter in your chart the information in the problem that refers to distance and its corresponding rate (speed).

- Fill in the **Distance Column** (*d*) with each distance given in the problem.
- Fill in the in the **Rate Column** (*r*) by naming the expressions for the rates as explained in 12.2.
- Fill in the **Time Column** (*t*) by using the variation of the distance formula to get the expressions.

  NOTE: Write the expression for the time as a **rational expression** (fraction) with distance over rate.

Step 3

Set Up An Equation

To set up your equation you will need the following information:

- The two rational expressions from the Time Column of the chart.
- The information given about time in the problem.
- Are the times the *same*? Or is the amount given the *total time*?

Motion – Rational Equations *Word Problem Workbook*

For the Motion Problems in this section, there are only two possible ways to set up an equation. This is determined by the information about time that is given in the problem.

- If the problem states that the **times are the same**, set the two rational expressions from the Time Column of the chart *equal to each other*.

- If a **total amount of time is given**,, *add together* the two rational expressions from the Time Column of the chart and set *equal to the total amount of time given*.

Time Given In Problem	Equation To Use
The Same	1st Rational Expression = 2nd Rational Expression
As A Total	1st Rational Expression + 2nd Rational Expressions = Total Time

Step 4
Solve the Equation

Using the method taught by your instructor, solve the equation for the variable.

Keep in mind when solving a Motion -- Rational Equations problem, you may get two solutions to the equation. If one of these solutions is negative, eliminate it because a rate (speed) **cannot be negative**.

> **NOTE**: If both of the solutions are positive, you must substitute each of them into the expressions named in Step 2. Whichever solution makes the value of any expression negative must be eliminated.

Step 5
Make Sure to Answer the Question Being Asked

In Motion Problems, as in other word problems, you need to make sure exactly what question is being asked. It is possible that the value for the variable x may be your answer. But it may *not* be.

For example, the value for x may be the rate of an object in still water, and the question may ask for the rate of the object if it is traveling upstream. In this case, you need to substitute the solution for the variable into the expression for an object traveling upstream that you named in Step 2.

EXAMPLES

EXAMPLE 1 When it was raining, Elliott drove for 120 miles. When the rain stopped, he drove 20 mph faster than he did while it was raining. He drove for 300 miles after the rain stopped. If Elliott drove for a total of 10 hours, how fast did he drive while it was raining?

SOLUTION

Step 1 *Read The Problem*

- There are two distances given. One is 120 miles and the other is 300 miles.
- There are Direct Translation words [faster than] in order to determine the rates.
- The rates are x, and $x + 20$. Fill them in next to their corresponding distances.
- The *total* time is given.

Step 2 *Fill In The Chart*

- In your chart, fill in the two distances given in the problem.
- Fill in the two expressions for the rates next to their corresponding distances.
- Fill in the Time by using the rational expression of Distance over Rate.

d	r	t
120	x	$\dfrac{120}{x}$
300	$x + 20$	$\dfrac{300}{x+20}$

Step 3 *Set Up The Equation*

- Use the two rational expressions for the time that you named in your chart in Step 2.
- The total time is given in the problem. It is 10 hours.
- Set up your equation as 1st rational expression + 2nd rational expression = total time.

$$\boxed{\dfrac{120}{x} + \dfrac{300}{x+20} = 10}$$

Motion – Rational Equations   Word Problem Workbook

Step 4 *Solve The Equation*

- The solution to the equation is $\boxed{30}$

Step 5 *Answer The Question Asked*

- You have the solution to the equation. The value of *x* is 30.
- 30 mph is the speed that Elliott drove for 120 miles while it was raining.
- The question asks the speed while raining. You are done. You have the correct answer.

    Answer: Elliott drove at a speed of 30 mph while it was raining.

EXAMPLE 2 Melody can cycle for 30 miles against the wind in the same amount of time that it takes her to cycle 66 miles with the wind. If the speed of the wind is 3 mph, what is Melody's speed when she cycles with the wind?

SOLUTION

Step 1 *Read The Problem*

- There are two distances given. One is 30 miles and the other is 66 miles.
- There is no Direct Translation.
- You are given the speed of the wind to determine the rates.
- As per 12.2, the rates are *x* – 3 against the wind, and *x* + 3 with the wind.
- The problem states the amount of time is the same.

Step 2 *Fill In The Chart*

- In your chart, fill in the two distances given in the problem.
- Fill in the two expressions for the rates next to their corresponding distances.
- Fill in the Time by using the rational expression of distance over rate.

Traveling	*d* (distance)	*r* (rate)	*t* (time)
Cycle against the wind	30	$x - 3$	$\dfrac{30}{x-3}$
Cycle with the wind	66	$x + 3$	$\dfrac{66}{x+3}$

Word Problem Workbook … Motion – Rational Equations

Step 3 *Set Up The Equation*

- Use the two rational expressions for the time that you named in your chart in Step 2.
- The problem states a total time in the problem. It is 8 hours.
- Set up your equation as 1st rational expression + 2nd rational expression = total time.

$$\frac{30}{x-3} = \frac{66}{x+3}$$

Step 4 *Solve The Equation*

- The solution to the equation is $x = 8$

Step 5 *Answer The Question Asked*

- You have the solution to the equation, but it is not the answer to the question.
- The value of x is 8, which is Melody's rate without the influence of the wind.
- The question asks for Melody's rate with the wind.
- Substitute 8 for x into the expression in your chart for her rate *with* the wind.

Rate with wind $= x + 3$
Rate with wind $= 8 + 3$
Rate with wind $= 11$

Answer: Melody's speed with the wind was 11 mph.

Section 12: Motion – Rational Equations Exercise Set

1. Paul can paddle a canoe 15 miles upstream in the same amount of time it takes Liz to paddle a canoe 27 miles downstream. If the current is 2 mph, what is Paul's speed as she travels downstream?

2. A cruise ship traveled for 275 miles with the current in the same amount of time it traveled 175 miles against the current. The speed of the current was 10 mph. What was the speed of the cruise ship while it traveled against the current?

3. Terry can run 15 miles in the same amount of time it takes Mark to run 21 miles. If Mark runs 2 mph faster than Terry, how fast does Mark run?

4. Doni can drive 100 miles to work in the same amount of time that it takes her to drive 60 miles to a friend's house. When driving to her friend's house, Doni drives 20 mph slower than when she drives to work. How fast does she drive going to work?

5. A boat travels 165 miles downstream in the same time the boat travels 135 miles upstream. If the speed of the current is 5 mph, what would be the speed of the boat in still water?

6. A steamboat travels 246 miles upstream in the amount of time the steamboat travels 294 miles downstream. If the speed of the current is 4 mph, how fast would the steamboat travel in still water?

7. The Goodyear blimp flies 153 miles with a tailwind in the same time it travels 57 miles with a headwind. If the speed of the wind is 16 mph, what is the speed of the blimp in still air?

8. A plane flies 960 miles with the wind in the same amount of time that the plane flies 640 miles against the wind. If the speed of the wind is 40 mph, what would the speed of the plane be in still air?

9. A two-engine Cessna flew for 510 miles with a tailwind of 40 mph in the same amount of time that it flew for 330 miles with a headwind of 20 mph. What was the speed of the Cessna when it was traveling with a headwind?

10. On their vacation, a family traveled 135 miles by train and then traveled 855 miles by plane. The speed of the plane was three times the speed of the train. If the total time of the trip was 6 hours, what was the speed of the train?

11. Harriet drove for 90 miles in the city. When she got on the highway she increased her speed by 20 mph and drove for 130 miles. If Harriet drove for a total of 4 hours, how fast did she drive on the highway?

12. Frankie Kowalski drove his jeep for 135 miles on a paved road. He decreased his speed by 25 mph when he went on a dirt rode. He traveled for 40 miles on the dirt rode. If Frankie drove his jeep for 5 hours, what was his speed when driving on the paved road?

13. A homing pigeon can fly 90 miles in the same amount of time an eagle can fly 100 miles. A pigeon flies 5 mph slower than an eagle. How fast can each bird fly?

14. During a marathon, Allen jogged 35 miles. When he got tired, he walked 6 miles to the finish line. Allen jogs 4 mph faster than he walks. If it took him 7 hours to finish the marathon, how fast was he jogging?

15. A hot air balloon travels 40 miles against the wind in the same amount of time that it travels 200 miles with the wind. If the speed of the wind is 20 mph, how fast does the hot air balloon travel against the wind?

16. Tom traveled 85 miles downstream on a raft in the same amount of time he traveled 15 miles upstream on his raft. If the speed of the current was 7 mph, how fast did he travel upstream?

17. Trevor rides a motorcycle 165 miles in the same amount of time he rides a bicycle 45 miles. Trevor can ride his motorcycle 40 mph faster than his bicycle. How fast does he ride his motorcycle?

18. A passenger train can travel 280 miles in the same amount of time a freight train can travel 140 miles. The passenger train travels 35 mph faster than the freight train. How fast does the passenger train travel?

19. On an island excursion, some tourists walked 7 miles on a nature path and then hiked 12 miles up a mountainside. The tourists walked 3 mph faster than they hiked. The total time of the excursion was 4 hours. At what rate did the tourists hike?

20. Karl is training for a triathlon. He trains for the swimming portion at the local beach One day Karl swam for a total of 6 hours. He swam 3 miles against the current, and 33 miles with the current. That day the speed of the current was 5 miles per hour. How fast was Karl swimming with the current?

Section 13

Work

There are two basic types of work problems. One type involves two people (or any two objects) **working together to complete a job**. The other type deals with a container that is being both **filled and emptied at the same time**.

Step 1
Read Through The Entire Problem

First you need to determine which type of work problem it is in order to know which equation setup to use.

- If the problem involves two people (or any two objects) working together to complete a job, that information will be clearly stated in the problem. You need to look for and take note of the time it takes **each person working alone** to complete the job, and how long it takes them **if they work together.**

- If the problem is the type that deals with a container that is being filled and emptied, that will be also be clearly stated in the problem. With this second type of work problem, you need to look for and take note of the **rate of time it takes to fill** the container and the **rate of time it takes to empty** the container.

Step 2
Naming The Expressions

When the Work Problem involves two people (or any two objects) working together to complete a job, your equation is set up using **two rational expressions** (fractions). Each fraction will be "**time together**" over "**time alone**".

- The "**time together**" is the time it takes the **two people working together** to complete the job. It may be given as an amount or it may be an unknown. Either way, it will always be *the same in both fractions*.

- The "**time alone**" is the amount of time it takes **one person working alone** to finish the job. The "time alone" will be *different for each fraction*, and will be either two given amounts, one given amount and one unknown, or two unknowns that you will name expressions for using Direct Translation.

When the work problem is the type that deals with a container that is being filled and emptied, you also set up two fractions.

- The first fraction will be the "**time together**" over the "**filling time**".
- The second fraction will be the "**time together**" over the "**emptying time**".
- The "filling time" and the "emptying time" will be given in the problem.
- The "**time together**" represents the total amount of time it takes to fill a container when it is being filled and emptied *at the same time*.
- The "time together" will always be the *same for both fractions*, and will always be an *unknown*.

Two People Working Together	"*Time Together*" is the amount of time it takes both people working together to complete the job.
Filling & Emptying A Container	"*Time Together*" is the amount of time it takes to fill the container when the container is being filled and emptied at the same time.

Step 3

Set Up An Equation

When the work problem is the type that has two people (or any two objects) working together to complete a job, each fraction represents the portion of the job that one person has completed. The equation you use is the **two fractions added together and set equal to "1"**. (The number "1" is used because the fractions represent each person's part of the job and the number "1" represents one whole job completed.)

When the work problem is the type that involves a container that is being filled and emptied, the first fraction represents the part when the container is being filled and the second fraction represents the part when the container is being emptied. To set up the equation, the **2nd fraction (the "emptying time") is *subtracted* from the 1st fraction (the "filling time") and set equal to "1"**. (The number "1" is used because the "filling time" fraction less the "emptying time" fraction represents one whole task completed.)

Type Of Work Problem	Equation To Use
Two People (Objects) Working Together	$\dfrac{\text{time together}}{\text{time alone}} + \dfrac{\text{time together}}{\text{time alone}} = 1$
Filling And Emptying A Container	$\dfrac{\text{time together}}{\text{filling time}} - \dfrac{\text{time together}}{\text{emptying time}} = 1$

Step 4
Solve the Equation

Using the method taught by your instructor, solve the equation for the variable. Keep in mind when solving a Work Problem, you may get two solutions to the equation. If one of these solutions is negative, eliminate it because an amount of time **cannot be negative**.

Step 5
Make Sure to Answer the Question Being Asked

You need to make sure what question is being asked in the problem. It is possible that the value for the variable x may be your answer. But it may *not* be.

The solution to the equation will be the answer to the question *if there is no Direct Translation used* to name expressions. If Direct Translation is used, you need to make sure of what the question is asking. To get the answer, you may need to substitute the value of x into the expression you named in Step 2.

Word Problem Workbook   Work

EXAMPLES

EXAMPLE 1 Steve can decorate a classroom in 3 hours. It takes Joel 4 hours to decorate the same classroom. How long will it take them to decorate the classroom working together?

SOLUTION

Step 1 *Read Through The Entire Problem*

- The problem involves two people working together.
- Steve can complete the job in 3 hours, working alone.
- Joel can complete the job in 4 hours, working alone.
- The time it takes them working together to complete the job is unknown.

Step 2 *Name The Expressions*

- The fractions will be "time together" over "time alone".
- The "time together" is unknown and will be represented by x in both fractions.
- The "time alone" for the 1st fraction representing Steve's time is given. It is 3.
- The "time alone" for the 2nd fraction representing Joel's time is given. It is 4.

Step 3 *Set Up The Equation*

- The equation used is 1st fraction plus 2nd fraction equals 1.
- The 1st fraction representing Steve's time is x over 3.
- The 2nd fraction representing Joel's time is x over 4.

$$\frac{x}{3} + \frac{x}{4} = 1$$

Step 4 *Solve The Equation*

- The solution to the equation is $x = 1\frac{5}{7}$

Step 5 *Answer The Question Asked*

- You have the solution to the equation.
- The question asks the time it takes both of them to complete the job working together.
- The value of x is the time it takes them to complete the job working together.
- You are done. You have the correct answer.

Answer: It takes them $1\frac{5}{7}$ hours to decorate the classroom working together.

EXAMPLE 2 The time it takes Elena to clean the kitchen is twice as long as the time it takes Camelia. Working together, they can clean the kitchen in 2 hours. How long would it take Elena to clean the kitchen working by herself?

SOLUTION

Step 1 *Read Through The Entire Problem*

- The problem involves two people working together.
- Elena takes twice the amount of time to complete the job as Camelia.
- The time it takes them working together to complete the job is 2 hours.

Step 2 *Name The Expressions*

- The fractions will be "time together" over "time alone".
- The "time together" is given. It is 2, and will be used in the numerator for both fractions.
- Use Direct Translation to determine the expressions for "time alone" for each fraction.
- The "time alone" it takes Camelia to do the work is unknown. Use the variable x.
- The "time alone" it takes Elena to do the work is twice as long as Camelia. Use $2x$.

Step 3 *Set Up The Equation*

- The equation used is 1st fraction plus 2nd fraction equals 1.
- The 1st fraction representing Camelia's time is 2 over x.
- The 2nd fraction representing Elena's time is 2 over $2x$.

$$\boxed{\frac{2}{x} + \frac{2}{2x} = 1}$$

Step 4 *Solve The Equation*

- The solution to the equation is $\boxed{x = 3}$

Step 5 *Answer The Question Asked*

- You have the solution to the equation, but it is not the answer.
- The question asks the time it takes Elena to complete the job working alone.
- The value of x is the time it takes Camelia to complete the job working alone.
- Substitute the solution for x (3) into the expression you named for Elena's time.

$$\boxed{\begin{array}{l} \text{Elena's Time} = 2x \\ \text{Elena's Time} = 2(3) \\ \text{Elena's Time} = 6 \end{array}}$$

Answer: It takes Elena 6 hours by herself.

EXAMPLE 3 Using a garden hose, it takes 15 minutes to fill up a child's wading pool. It takes 45 minutes to empty out the pool through a small drainage hole. If the drainage hole is left open by mistake, and the water is draining out at the same time the pool is being filled with the garden hose, how long will it take before the pool is filled up?

SOLUTION

Step 1 *Read Through The Entire Problem*

- The problem involves a container being filled and emptied.
- It takes 15 minutes to fill up the pool.
- It takes 45 minutes to empty out the pool.

Step 2 *Name The Expressions*

- The 1st fraction will be "time together" over "filling time".
- The 2nd fraction will be "time together" over "emptying time".
- The "time together" is always unknown. Use x as the numerator for both fractions.
- The "filling time" of the 1st fraction is given. It is 15 minutes.
- The "emptying time" of the 2nd fraction is given. It is 45 minutes.

Step 3 *Set Up The Equation*

- The equation used is 1st fraction (filling) minus 2nd fraction (emptying) equals 1.
- The 1st fraction representing the filling time is x over 15.
- The 2nd fraction representing emptying time is x over 45.

$$\boxed{\frac{x}{15} - \frac{x}{45} = 1}$$

Step 4 *Solve The Equation*

- The solution to the equation is $\boxed{x = 22\frac{1}{2}}$

Step 5 *Answer The Question Asked*

- You have the solution to the equation.
- There is no Direct Translation used to name the expressions.
- The solution to the equation is the answer to the question that is asked.
- You are done. You have the correct answer.

Answer: It will take $22\frac{1}{2}$ minutes to fill up the pool.

Section 13: Work Exercise Set

1. Beavis can wash the walls in 3 hours working alone, and Bart can wash the walls in 2 hours. How long would it take them to wash the walls if they worked together?

2. Eric can install the carpet in a room in 3 hours, but Mark needs 5 hours. How long will it take them to complete the job if they work together?

3. Working alone, it takes Jacob 5 minutes longer to wash the dishes than it takes Sarah when she does it alone. Washing the dishes together, Jacob and Sarah can finish the dishes in 6 minutes. How long does it take Jacob to wash the dishes by himself?

4. It takes Vinnie 2 hours to groom a poodle by himself. Marie can groom the same poodle in 4 hours. How long would it take them if they worked together to groom the poodle?

5. Working together, Benjamin and Alex can paint a room in 4 hours. Working alone, it would take Benjamin 7 hours to paint the room. How long would it take Alex to paint the room working alone?

6. Working together, Michelle and Lauren can make a quilt in 5 days. If it takes Michelle 8 days to make a quilt by herself, how long would it take Lauren to make a quilt by herself?

7. It takes an older inkjet printer three times as long as it takes a new laser printer to print out a math workbook. With both printers running, the workbook can be printed out in 9 hours. How long would it take the old printer to print out the whole workbook by itself?

8. It takes Jarred 3 minutes longer than Sierra to eat a chocolate bar. If it takes them 2 minutes to eat the same chocolate bar together, how long does it take Jarred to eat the chocolate bar by himself?

9. It takes Luke 9 hours longer than Laura to do the store's inventory. If they can finish the inventory in 6 hours working together, how long does it take Laura to do the inventory alone?

10. It takes Ashley 3 times longer than Taylor to landscape their backyard. Working together, they can they can finish the job in 21 days. How long does it take Ashley to landscape the backyard if she works alone?

11. It takes Mr. Spock 30 minutes to fight off a Klingon warship by himself. Captain Kirk can fight them off in 45 minutes. How long will it take them to fight off the Klingons if they fight together?

12. Don can write a research paper in 7 days. His girlfriend, Adrienne, can write a research paper in 4 days. If the instructor assigned this as a team project, how long would it take Don and Adrienne if they write the research paper together?

13. Ramses and Moses could build a pyramid in 3 months. If it takes Ramses 5 months to build a pyramid by himself, how long would it take Moses to build a pyramid by himself?

14. It takes Jerry 18 more minutes than Dean to wash a car. If they can wash the car together in 12 minutes, how long does it take Jerry to wash the car by himself?

15. On Nickelodeon's Double Dare Show, a large vat of green goop can be filled by a hose in 12 hours. The vat can be emptied by a different hose in 15 hours. How long will it take to fill the vat with goop if both the filling hose and the emptying hose are both being used?

16. It takes 9 minutes to fill up the kitchen sink with water from the faucet, and it takes 12 minutes to drain the water out. How long will it take to fill up the sink if the drain is left open.

17. It takes 10 minutes to fill a water cooler with filtered water and 12 minutes to empty the water cooler with the drain plug open. Donovan tried to fill up the water cooler without realizing that the drain plug had been left open. How long did it take Donovan to fill up the water cooler?

18. It takes 20 hours to fill up a cement-mixing truck with cement and 25 hours to empty out the truck. If the cement is pouring out of the truck at the same time it is being poured into the truck, how long will it take to completely fill the truck with cement?

19. One pipe can fill an oil tanker in 5 hours and another pipe can empty the oil tanker in 10 hours. If both pipes are in use at the same time, how long will it take to fill up the oil tanker?

20. It takes 4 minutes to fill up a bucket of paint. If someone pokes a large hole in the bottom of the bucket, the paint will empty out in 6 minutes. If a painter tries to fill up this bucket with the hole in the bottom, how long will it take him?

Answers

Section 1 1. $12 + x = 10$ 3. $2x - 2 = 11$ 5. $x^3 - 6 = 15$ 7. $5(6 + x) = 24$ 9. $4x - 3 = 5x + 10$ 11. $4(x + 10) = -92$ 13. $94 - x = 5x + 19$ 15. $9x - 8 = 6 + x^2$ 17. $11(x - 42) = x^3 + 10$ 19. $x^2 - 7 = 2(5x + 2)$

Section 2 1. 4   3. 700   5. 20%   7. 12   9. 60   11. 22%   13. 4.81   15. 10   17. 500%   19. 18

Section 3 1. 8 tanks of water   3. $28\frac{4}{5}$ acres   5. 187 points   7. 360 points   9. 625 yards   11. $126.50   13. 216 grams of carbohydrates   15. 100 pounds   17. 30 reruns   19. 120 minutes

Section 4 1. 43 years old   3. $33   5. 20 pieces of milk chocolate; 8 pieces of dark chocolate   7. lowest grade = 19; highest grade = 76   9. 510 women   11. 32 feet   13. 824,346 votes   15. 14 feet   17. 4 inches, 11 inches, 22 inches   19. 4643 Democrats

Section 5 1. 12 meters   3. 17 inches   5. 14 centimeters   7. 7 yards   9. width = 11 feet; length = 18 feet   11. 450 feet   13. 13 feet   15. 15 inches   17. width = 37 cm; length = 47 cm   19. width = 7 feet; length = 28 feet

Section 6 1. 20, 21   3. 15, 17   5. 32, 34   7. 30, 31, 32   9. 23, 25, 27   11. 22, 24, 26   13. 15   15. 11   17. 52   19. 6

Section 7 1. 2 hours   3. 2 hours   5. 3 hours   7. 3 hours   9. $2\frac{1}{2}$ hours   11. 1 hour   13. 10 hours   15. 17 mph   17. 10 mph; 25 mph   19. 20 mph

Section 8 1. 13 quarters   3. 10 five-dollar bills   5. 50 dimes; 122 nickels   7. 17 five-dollar bills; 18 ten-dollar bills   9. 24 dimes; 22 quarters   11. 12 fifties; 28 twenties   13. 3 fifties   15. 520 pennies; 100 dimes   17. 12 twenties   19. 22 ten-dollar bills

Section 9 1. $18,500   3. $6000   5. $7000 at 9%; $6000 at 11%   7. $1500   9. $400 at 9%; $650 at 10%   11. $1500   13. $5000   15. $7000 at 6.5%; $5000 at 4.5%   17. $2000 at 6%; $2500 at 8%   19. $45,000

Section 10 1. 32 liters   3. 25 pounds   5. 7.5 liters   7. 84 milliliters   9. 4 oz   11. 30 pints   13. 12 cubic inches   15. 40 tablespoons   17. 6 cups   19. 150 oz

Section 11 1. width = 4 inches; length = 11 inches   3. 11 feet   5. 5 inches by 10 inches   7. 3 yards by 5 yards   9. 10 feet   11. 5 feet, 12 feet, 13 feet   13. 6 inches   15. 12 meters, 16 meters, 20 meters   17. 7 yards, 24 yards, 25 yards   19. 6 feet

Section 12 1. 5 mph   3. 7 mph   5. 50 mph   7. 35 mph   9. 110 mph   11. 65 mph   13. pigeon = 45 mph; eagle = 50 mph   15. 10 mph   17. 55 mph   19. 4 mph

Section 13 1. $1\frac{1}{5}$ hours   3. 15 minutes   5. $9\frac{1}{3}$ hours   7. 36 hours   9. 9 hours   11. 18 minutes   13. $7\frac{1}{2}$ months   15. 60 hours   17. 60 minutes   19. 10 hours